SpringerBriefs in Electrical and Computer Engineering

Control, Automation and Robotics

AF182461

For further volumes:
http://www.springer.com/series/10198

Giovanni Fiengo · Alessandro di Gaeta
Angelo Palladino · Veniero Giglio

Common Rail System for GDI Engines

Modelling, Identification, and Control

Springer

Giovanni Fiengo
Dipartimento di Ingegneria
Università degli studi del Sannio
Benevento
Italy

Alessandro di Gaeta
Istituto Motori
Consiglio Nazionale delle Ricerche
Naples
Italy

Angelo Palladino
Dipartimento di Ingegneria
Università degli studi del Sannio
Benevento
Italy

Veniero Giglio
National Research Council
Istituto Motori
Napoli
Italy

ISSN 2192-6786 ISSN 2192-6794 (electronic)
ISBN 978-1-4471-4467-0 ISBN 978-1-4471-4468-7 (eBook)
DOI 10.1007/978-1-4471-4468-7
Springer London Heidelberg New York Dordrecht

Library of Congress Control Number: 2012944967

© The Author(s) 2013
This work is subject to copyright. All rights are reserved by the Publisher, whether the whole or part of the material is concerned, specifically the rights of translation, reprinting, reuse of illustrations, recitation, broadcasting, reproduction on microfilms or in any other physical way, and transmission or information storage and retrieval, electronic adaptation, computer software, or by similar or dissimilar methodology now known or hereafter developed. Exempted from this legal reservation are brief excerpts in connection with reviews or scholarly analysis or material supplied specifically for the purpose of being entered and executed on a computer system, for exclusive use by the purchaser of the work. Duplication of this publication or parts thereof is permitted only under the provisions of the Copyright Law of the Publisher's location, in its current version, and permission for use must always be obtained from Springer. Permissions for use may be obtained through RightsLink at the Copyright Clearance Center. Violations are liable to prosecution under the respective Copyright Law.
The use of general descriptive names, registered names, trademarks, service marks, etc. in this publication does not imply, even in the absence of a specific statement, that such names are exempt from the relevant protective laws and regulations and therefore free for general use.
While the advice and information in this book are believed to be true and accurate at the date of publication, neither the authors nor the editors nor the publisher can accept any legal responsibility for any errors or omissions that may be made. The publisher makes no warranty, express or implied, with respect to the material contained herein.

Printed on acid-free paper

Springer is part of Springer Science+Business Media (www.springer.com)

Contents

About the Editors

Tamer Başar is with the University of Illinois at Urbana-Champaign, where he holds the academic positions of Swanlund Endowed Chair, Center for Advanced Study Professor of Electrical and Computer Engineering, Research Professor at the Coordinated Science Laboratory, and Research Professor at the Information Trust Institute. He received his B.S.E.E. degree from Robert College, Istanbul, and the M.S., M.Phil, and Ph.D. degrees from Yale University. He has published extensively in systems, control, communications, and dynamic games, and has current research interests that address fundamental issues in these areas along with applications such as formation in adversarial environments, network security, resilience in cyber-physical systems, and pricing in networks.

In addition to his editorial involvement with these *Briefs*, Başar is also the Editor-in-Chief of *Automatica*, Editor of two Birkhäuser Series on *Systems & Control* and *Static & Dynamic Game Theory,* the Managing Editor of the *Annals of the International Society of Dynamic Games* (ISDG), and member of editorial and advisory boards of several international journals in control, wireless networks, and applied mathematics. He has received several awards and recognitions over the years, among which are the Medal of Science of Turkey (1993); Bode Lecture Prize (2004) of IEEE CSS; Quazza Medal (2005) of IFAC; Bellman Control Heritage Award (2006) of AACC; and Isaacs Award (2010) of ISDG. He is a member of the US National Academy of Engineering, Fellow of IEEE and IFAC, Council Member of IFAC (2011–2014), a past president of CSS, the founding president of ISDG, and president of AACC (2010–2011).

Antonio Bicchi is Professor of Automatic Control and Robotics at the University of Pisa. He graduated at the University of Bologna in 1988 and was a postdoc scholar at M.I.T.A.I. Lab between 1988 and 1990. His main research interests are in:

- Dynamics, kinematics and control of complex mechanical systems, including robots, autonomous vehicles, and automotive systems;
- Haptics and dextrous manipulation; and

- Theory and control of nonlinear systems, in particular hybrid (logic/dynamic, symbol/signal) systems.

He has published over 300 papers on international journals, books, and refereed conferences. Professor Bicchi currently serves as the Director of the Interdepartmental Research Center "E. Piaggio" of the University of Pisa, and President of the Italian Association or Researchers in Automatic Control. He has served as Editor-in-Chief of the Conference Editorial Board for the IEEE Robotics and Automation Society (RAS), and as Vice President of IEEE RAS, Distinguished Lecturer, and Editor for several scientific journals including the *International Journal of Robotics Research*, the *IEEE Transactions on Robotics and Automation*, and *IEEE RAS Magazine*. He has organized and co-chaired the first World Haptics Conference (2005), and Hybrid Systems: Computation and Control (2007). He is the recipient of several best paper awards at various conferences, and of an Advanced Grant from the European Research Council. Antonio Bicchi has been an IEEE Fellow since 2005.

Miroslav Krstic holds the Daniel L. Alspach chair and is the founding director of the Cymer Center for Control Systems and Dynamics at University of California, San Diego. He is a recipient of the PECASE, NSF Career, and ONR Young Investigator Awards, as well as the Axelby and Schuck Paper Prizes. Professor Krstic was the first recipient of the UCSD Research Award in the area of engineering and has held the Russell Severance Springer Distinguished Visiting Professorship at UC Berkeley and the Harold W. Sorenson Distinguished Professorship at UCSD. He is a Fellow of IEEE and IFAC. Professor Krstic serves as the Senior Editor for *Automatica* and *IEEE Transactions on Automatic Control* and as an Editor for the Springer series *Communications and Control Engineering*. He has served as Vice President for Technical Activities of the IEEE Control Systems Society. Krstic has co-authored eight books on adaptive, nonlinear, and stochastic control, extremum seeking, control of PDE systems including turbulent flows, and control of delay systems.

Chapter 1
Introduction to Internal Combustion Engines

Abstract The Internal Combustion Engine (ICE) is the technological innovation that has changed the world. It is considered both as one of the greater sources of benefits and one of the main reasons of the atmospheric pollution. Therefore, in order to satisfy polluting emission legislations with particular attention for CO_2 production, the engine design has been continuously improved and new components have been introduced, such as sensors and complete subsystems. The generation of polluting substances produced by the combustion depends not only on the engine structure, but also on the engine management system that is mainly responsible for the engine behavior, and so for the emission levels. Because of the introduction of new electronic components, which determines a strong increase of the degrees of freedom available during the engine operations, the optimization of engine behavior becomes more complex and less intuitive thus not obtaining the best performances of the system. Nowadays, the engine control system is an enormous collection of tuning tables, obtained adding the new tables related to new electronic components to the previous ones. This control structure is usually defined using trial-and-error approaches: engineers dedicated to the calibration process transform the goals, such as fuel consumption, engine emission, performances of the vehicle, and so on, in control references to be followed by some engine and vehicle variables that can be measured on-line. These goals are obtained through proper closed-loop control strategies receiving measurements by sensors and acting on actuators. In this chapter after a short description of the operating principles of reciprocating ICEs and their classification, an overview of a functional architecture of an Engine Management System (EMS) is given.

1.1 Basic Concepts of Internal Combustion Engines

The design and development of ICE has encouraged for long time a number of research activities that can be in general split into two main directions: the first is essentially devoted to the analysis of existing engines aiming at control and

G. Fiengo et al., *Common Rail System for GDI Engines*, SpringerBriefs in Control, Automation and Robotics, DOI: 10.1007/978-1-4471-4468-7_1, © The Author(s) 2013

performance optimization in various working conditions; the second has the main purpose of designing a novel engine by defining the configuration and its main geometric variables. During the previous years, an essential experimental approach was predominant in both these research fields. This is mainly due to the complexity of physical phenomena occurring within an ICE and therefore to a very difficult theoretical analysis. Actually, the steady-state operations of a reciprocating engine is not stationary but periodic, and equations describing the mass flow within the pipes are nonlinear, therefore they have no analytic solutions (the reader is referred to specialized textbooks as [1–4]).

Moreover, in spite of the increased knowledge on the physics describing turbulence phenomena, combustion, and fuel injection, research activities on ICE maintain at present an essential experimental character. Referring to the analysis of existing engines, for example, it is well-known that, for each new prototype, as well as subsequent development of engines already manufactured, it is essential, aiming at designing the engine electronic control strategies, to carry out a great number of experimental tests in order to define the optimal combination of all control parameters for all engine working conditions.

The progressive reduction in vehicle emission limits required the automotive industry to invest in research for developing alternative and more efficient control strategies. All control features and resources are permanently active in an Electronic Control Unit (ECU), ensuring the best performance with respect to emission, fuel economy, driveability and diagnostics, independently from engine working point. The aim of the ICE is the conversion of the chemical energy contained in the fuel into mechanical energy. The ICE differs from the external combustion engine, because the thermal energy is released by burning the fuel inside the engine. The air–fuel mixture before the combustion and the burned products after combustion are the actual working fluids. The work production, which provides the desired power output, occurs directly between these working fluids and the mechanical components of the engine.

According to the combustion mode, internal combustion engines can be classified as:

- *Volumetric or intermittent-combustion engine*: in these engines, air and fuel (the working fluid) are periodically trapped and ignited in the combustion chamber whose volume varies periodically thanks to the motion of some mechanical parts. In this family of propulsors there are reciprocating engines, in which the alternate motion of the piston is converted into a rotary motion through a connecting rod-crank linkage.
- *Continuous-combustion engine*: in these engines, the combustion occurs continuously thanks to a steady flow of air and fuel (producing the working fluid) in the combustion chamber whose volume is fixed. Gas turbines and jet engines are part of this family of propulsors.

As regards the volumetric internal combustion engines, it is possible to consider the further three classification methods:

- According to the *principle of operation*: the ICE can be 2-stroke or 4-stroke. In the case of 2-stroke engine, a cycle of operation (including the phases of exhaust and intake of the working fluid) is completed within one revolution of the crankshaft. In the case of a 4-stroke engine, the cycle requires two revolutions of the crankshaft.
- Depending on the *mode of combustion activation*: Spark Ignition (SI) or Compression Ignition (CI). In an SI engine, the spark actuation initiates the charge combustion process, which is of premixed type. In a CI (or diesel) engine, combustion is initiated by the conditions of pressure and temperature which are created in the combustion chamber during the compression phase, and the combustion process is termed diffusive.
- According to the *fuel delivery*: indirect or Port Fuel Injection (PFI) and Direct Injection (DI) engines. In the case of a PFI engine, the fuel is injected upstream of the combustion chamber, within the intake pipe. Instead, for the DI engine, the fuel is directly injected into the combustion chamber.

1.1.1 Operation of Reciprocating Engines

The moving part of a reciprocating ICE which exchanges work with gas is the piston, mostly of circular cross-section. The piston moves within the cylinder between two extreme positions, said dead centers. In the position of Top Dead Center (TDC), the piston identifies the minimum in-cylinder volume, V_2, which also constitutes the combustion chamber, while at the Bottom Dead Center (BDC) the maximum in-cylinder volume, V_1, is identified. It is also define:

- Stroke, $C = 2r$ is the distance between TDC and BDC and r is the crank radius;
- Bore, D is the diameter of the cylinder (if circular);
- Engine displacement, $V = V_1 - V_2 = \pi(D/2)^2 C$;
- Volumetric compression ratio, $\rho = V_1/V_2$ (Fig. 1.1).

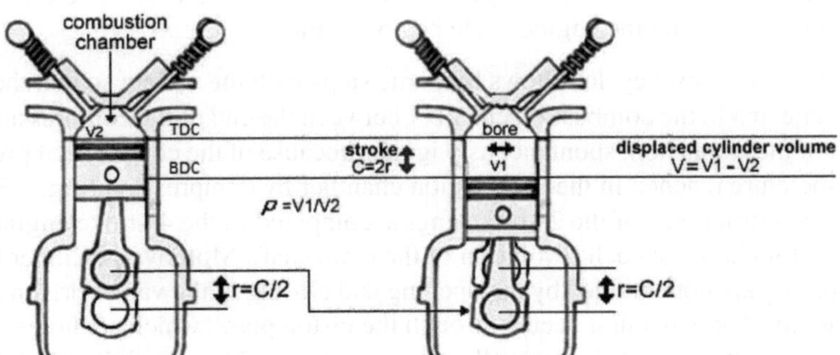

Fig. 1.1 Main geometric parameters of an ICE

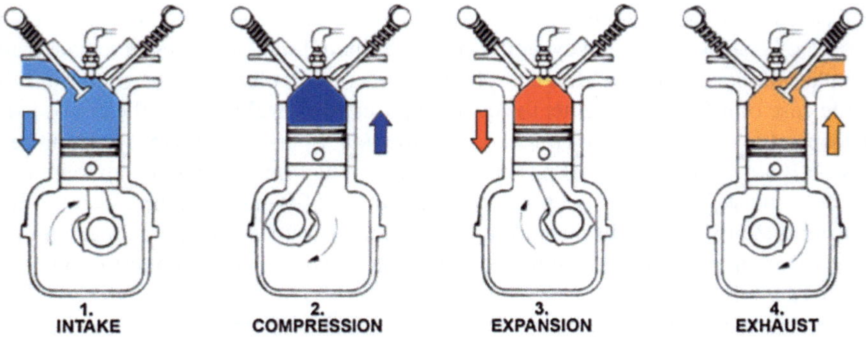

Fig. 1.2 Phases of a 4-stroke spark ignition engine

The 4-stroke engine is characterized by a useful phase every two revolutions of the engine, or every four strokes of the piston. Figure 1.2 illustrates the phases of the ideal cycle of a 4-stroke SI engine, detailed as follows:

- *Intake*: the fresh charge goes into the cylinder, through one or more valves, sucked by the vacuum created by the motion of the piston that moves from TDC to BDC.
- *Compression*: the intake valve closes when the piston reaches the BDC and the compression phase starts. The fresh charge is compressed by the upward motion of the piston. The in-cylinder volume is reduced from V_1 to V_2. At the end of the compression phase, a spark supplied by a special auxiliary body (Spark Plug) starts the ignition of the mixture. The combustion determines the raising of pressure and temperature inside the combustion chamber for the subsequent expansion.
- *Expansion*: is the only useful phase of the cycle, in which the fluid gives work to the piston that moves from TDC to BDC, where the opening of the exhaust valve occurs.
- *Exhaust*: the upward motion of the piston from BDC to TDC determines a dynamic action pushing the burnt gases out of the cylinder through the exhaust valve. When the piston is at TDC the engine cycle is completed, the exhaust valve closes, the inlet port opens, and the engine cycle begins again.

The 4-stroke Diesel cycle follows the same steps with the variant so that the fuel, directly injected in the combustion chamber between the end of the compression and the start of the expansion, spontaneously ignites because of the condition of pressure and temperature reached in the combustion chamber by compression (Fig. 1.3).

The main difference of the 2-stroke engine compared to the 4-stroke engine is to have a useful phase for each revolution of the crankshaft. Moreover, cylinder filling and emptying are not obtained by the opening and closing of the valves driven by the valvetrain mechanism, but it occurs through the piston ports, which are holes placed circumferentially on the cylinder wall, opened and closed by the piston skirt during its own motion in the cylinder.

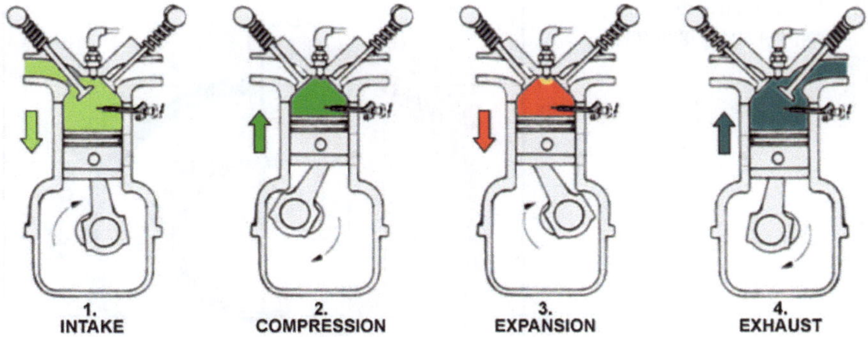

Fig. 1.3 Phases of a 4-stroke diesel engine

Fig. 1.4 Phases of a 2-stroke engine

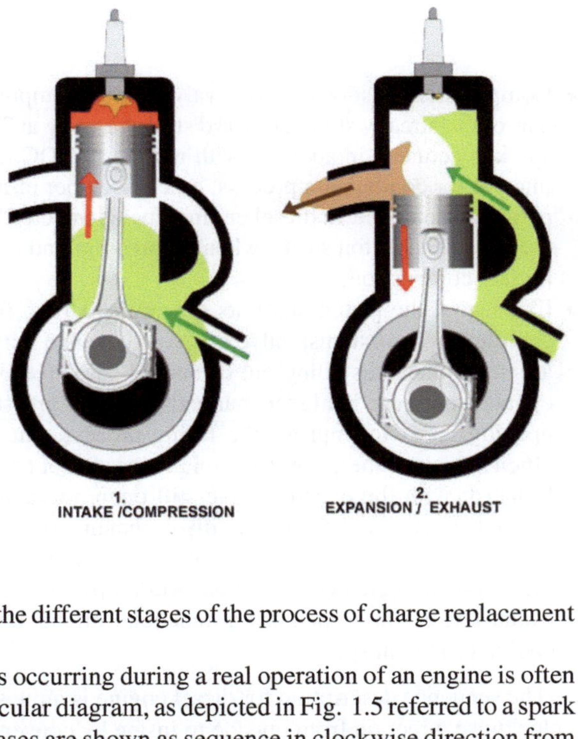

Figure 1.4 shows in detail the different stages of the process of charge replacement in a 2-stroke ICE.

The sequence of the phases occurring during a real operation of an engine is often described through a sort of circular diagram, as depicted in Fig. 1.5 referred to a spark ignition engine, where the phases are shown as sequence in clockwise direction from inside to outside.

- Intake: the intake valve is opened in advance with respect to TDC, by an angle that can range up to 20°. After air induction during the downward stroke, the inlet valve is closed with an angle delay with respect to BDC, exploiting gas inertia to enhance cylinder filling and consequently the resulting work per cycle.

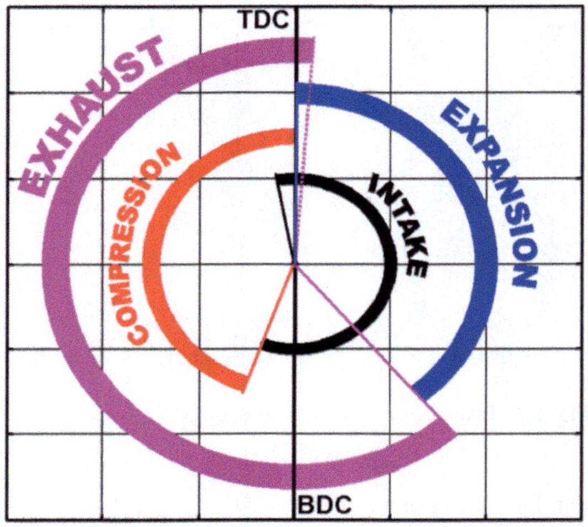

Fig. 1.5 Diagram of the phase sequence in a spark ignition engine

- Compression: as soon as intake valve closes, compression begins in the remaining part of the already started upward stroke, ending at TDC. In spark ignition engine, spark is actuated in advance with respect to TDC to guarantee an adequate time interval for combustion process, otherwise poor efficiency and high exhaust temperature can result. In diesel engine, fuel is injected before TDC reaching, toward the end of the piston stroke when air pressure and temperature are high enough to initiate combustion.
- Expansion: the piston starts its downward stroke from TDC, while combustion completes. The exhaust valve opens in advance with respect to BDC, promoting a first spontaneous going out of burned gases (blowdown) up to BDC, since in-cylinder pressure is higher than exhaust pipe pressure. This advance in exhaust opening is set to improve the filling process, but has some negative effect on efficiency since the gas expansion process is not completely exploited.
- Exhaust: with the exhaust valve still open, the piston starts the second upward stroke to scavenge the cylinder (displacement phase) of burned gases. The exhaust valve remains open up to several crank angle degrees past the TDC position; thus a crank angle period (called overlap period) results in which both inlet and exhaust valves are open, with beneficial effects on fresh charge filling at high engine rotational speed.

The sequence description for diesel engine is quite similar, with the difference of a clockwise direction from outside to inside by convention.

1.1.2 Main Engine Operating Parameters

In-cylinder pressure can be measured during the operating cycle of the engine, and represented in a pressure–volume diagram, as in Fig. 1.6, called *indicated pressure*

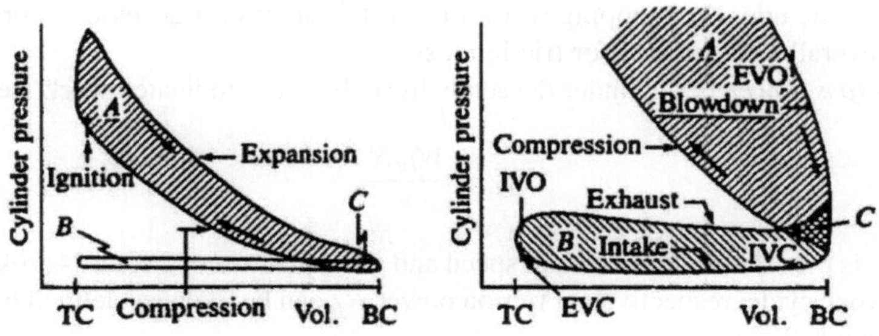

Fig. 1.6 Typical P–V diagram of a spark ignition engine (taken from [1])

diagram. This data can be used to calculate the work transferred from the gas to the piston, by integrating pressure versus volume along the closed curve, between volumes corresponding to TDC and BDC positions. The integral

$$W_i = \int p dV, \tag{1.1}$$

corresponding to the area enclosed in the curve, is called *indicated work per cycle*. For a 4-stroke engine, a double definition is used, differentiating the *gross indicated work per cycle* W_{ig} (that is the work calculated only along the compression and expansion phases, when both inlet and exhaust valves are closed, $\text{area}(A)+\text{area}(C)$ in Fig. 1.6) from the *net indicated work per cycle* W_{in} (that is the work calculated along the whole pressure cycle, corresponding to $[\text{area}(A)+\text{area}(C)]-[\text{area}(B)+\text{area}(C)]$).

The $\text{area}(B)+\text{area}(C)$ is called *pumping work*, being the work exchanged between in-cylinder gases and piston during the intake and exhaust phases. In a normally aspirated engine, this work is given from the piston to the gas, since in-cylinder pressure during the intake phase is lower than that in the exhaust phase. The contrary occurs in turbocharged engines with intake pressure higher than exhaust pressure. The ratio between the indicated work per cycle and engine displacement V, having units of force per unit area, is called Indicated Mean Effective Pressure (IMEP):

$$\text{IMEP} = \frac{W}{V}, \tag{1.2}$$

which corresponds to a constant pressure value such as, multiplied by engine displacement V, provides the work resulting from the integral (1.1). The gross indicated work per cycle yielded to the piston is decreased by the pumping losses and by the various friction losses occurring between the piston–ring assembly and the cylinder, along the connecting rod linkage mechanism to the crank shaft and engine flywheel. Other work is lost to operate the mechanisms used for valve actuation, water and oil pumps, and other accessories, and to overcome their friction losses. Thus, the work per cycle actually available W_b at engine flywheel is

$$W_b = W_{ig} - W_f, \tag{1.3}$$

where W_f includes the pumping work, the work required for accessories operation and the overall work wasted for friction losses.

The *gross power* per cylinder descends from the gross indicated work, i.e.,

$$P_{ig} = \frac{W_{ig}N}{\varepsilon}, \tag{1.4}$$

where N is the crankshaft rotational speed and ε is equal to 1 or 2 for a 2-stroke cycle or a 4-stroke cycle, respectively. Friction power P_f can be likewise defined based on friction work W_f as

$$P_f = \frac{W_f N}{\varepsilon}. \tag{1.5}$$

The actual power available at engine flywheel is measured coupling the engine to a dynamometer, and is called *brake power*. Based on the previous definitions, brake power is

$$P_b = P_{ig} - P_f. \tag{1.6}$$

Engine mechanical efficiency is defined as the ratio between the brake power and the gross indicated power:

$$\eta_m = \frac{P_b}{P_{ig}} \tag{1.7}$$

or, substituting Eq. (1.6) it has

$$\eta_m = 1 - \frac{P_f}{P_{ig}}. \tag{1.8}$$

Remembering that friction power includes pumping work, it results that this one affects negatively engine mechanical efficiency, giving the explanation of efficiency decay produced by low load throttling in spark ignition engines.

An operation variable often used is the Brake Mean Effective Pressure (BMEP), defined as:

$$\text{BMEP} = \frac{P_b \varepsilon}{VN}, \tag{1.9}$$

which is useful to identify engine load, as well as for design purposes to estimate engine displacement required for an assigned rated power.

1.2 Outline of the Different Types of Reciprocating Engines

1.2.1 PFI-SI Engines

In PFI-SI engines, the mixture of fuel and air is formed in the intake manifold. Here the fuel is atomized and vaporized, resulting in a homogenous mixture with air. The charge is then forced into the combustion chamber during the intake stroke and consequently compressed when the intake valves are closed and the piston starts to move up. Close to TDC, a spark discharge starts the ignition of the charge and the flame propagates through the cylinder until the walls where it is quenched.

SI engines usually works in conjunction with Three-Way Catalytic (TWC) conversion system which substantially decreases the engine emissions of carbon monoxide (CO), nitrogen oxides (NO_x), and unburned HydroCarbons (HC). To guarantee the maximum conversion efficiency of the TWC, the air and fuel mixture is kept close to stoichiometric proportions in almost all loads. The load of the engine is controlled by regulating the quantity of air mass crossing the throttle valve and then flowing into the engine.

In this type of engine, air throttling leads to an increase of pumping losses since the engine has to spend more work to induct the air/fuel mixture into the cylinder, resulting in a decreased efficiency [1]. The efficiency directly affects the fuel consumption, which results in higher carbon dioxide (CO_2) emissions.

1.2.2 GDI-SI Engines

In Gasoline Direct Injection (GDI) engines, the placement of the injectors has been moved from the intake manifold (as in PFI engines) to the combustion chamber, as shown in Fig. 1.7. Taking advantage of a special piston design with no-flat head and high-pressure fueling system, GDI engines can be operated in two distinct modes, i.e., *Homogenous Combustion Mode* and *Stratified Combustion Mode*, by varying the

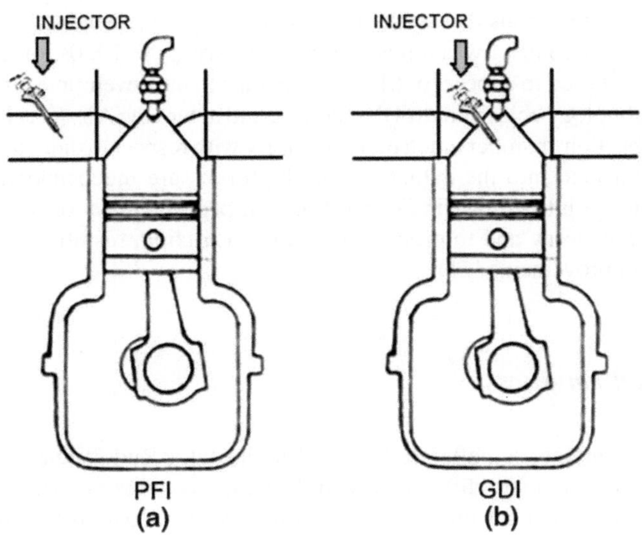

Fig. 1.7 Injector placement in (**a**) a PFI engine and (**b**) a GDI engine

fuel injection timing [5, 6]. The homogenous combustion mode is obtained injecting the fuel during the intake stroke. The engine performance, both in terms of torque and emissions, is comparable to those of conventional PFI engines since a homogeneous stoichiometric air/fuel mixture is used for the combustion. Instead, the stratified combustion mode is obtained injecting the fuel during the compression stroke, and the engine works with a stratified air–fuel mixture that is globally lean.

Several advantages can result [7] from the injection of fuel directly into the combustion chamber:

- Higher engine efficiency: the vaporization of the fuel in the cylinder causes a temperature decrease of the charge; a charge with a lower temperature is less prone to knock, which makes it possible to use an higher compression ratio leading to a higher engine efficiency.
- Reduction of pump losses: when the engine runs unthrottled, the power output is controlled by the amount of fuel injected into the cylinder, and not relying on a throttle valve as in conventional PFI engines, leading to lower pumping losses and thus higher efficiency.
- Reduction of fuel consumption: regulating the fuel injection during the compression stroke for charge stratification, it is possible to create around the spark plug a small region of stoichiometric mixture that can be easily ignited, while maintaining a lean mixture in the rest of the combustion chamber. On the whole, it results in a lower quantity of fuel injected.

The downsides are emissions of particulate matter and NO_x. Working with an ultra-lean mixture, if from one side it is possible to reach a very good fuel economy, conversely the normal TWC does not properly work for NO_x reduction and it is necessary to equip the engine with an additional de-NO_x catalyst/trap able to reduce also the particle emissions that are higher than those of a PFI engine.

Although GDI concept was already proposed during the 1920s, only in the last decades it has started to appear [6, 8]. This is thanks to the development of Common Rail (CR) technology able to control the fuel rate and in turn the Air–Fuel Ratio (AFR) into the combustion chamber much more precisely with respect to older technologies. The fuel is injected into the cylinders with high pressure independently from the engine working point as shown in [9, 10]. The high-pressure injection allows to finely atomize the fuel spray and to promote fuel and air mixing, resulting in significant combustion improvements [11].

1.2.3 Diesel Engines

Diesel (or CI) engine, so-called in honor of its inventor Rudolf Diesel, is a direct injected engine. The main difference with SI engine is the combustion process, in which the adopted fuel has the capacity to ignite spontaneously if the air inside the cylinder becomes hot enough through compression [1, 12]. Diesel engines have an efficiency higher than gasoline engines (45 % vs. 30 %).

Compression-ignited diesel engine has a number of advantages due to the nature of the combustion process:

- It does not need to throttle the intake air since it is operated with an inhomogeneous mixture, so that the load can be controlled by the amount of fuel injected into the cylinder. Hence, CI engine avoids pumping losses at part load which results in a higher mechanical efficiency.
- Thanks to the diffusive combustion process there are not the limits on the compression ratio imposed by the knock phenomenon on SI engines. Therefore, higher compression ratios (up to 20:1 vs. 13:1 for SI engines) are realizable resulting in higher thermodynamic efficiency since more work is extracted during the expansion stroke.
- The CI engine operates always lean so that the ratio of specific heats is higher over the expansion process which also results in a higher fuel conversion efficiency, and therefore low fuel consumption and lower emissions of CO_2.
- CI engine has higher volumetric efficiency since there are no flow losses at throttle, no intake manifold heating, no fuel vapor (since the fresh charge does not contain any fuel).
- Without the knock phenomenon, the cylinders of a diesel engine can be designed larger which leads to lower heat losses for engines of comparable power. This is due to the fact that engine power increases with displacement volume, but heat losses with surface area. Thus, large diesel engines (e.g., for marine propulsion) can achieve brake efficiencies around 50 %.

The drawback of low power density of naturally aspirated diesel engine is compensated equipping the Diesel engine with a turbocharger (turbine + compressor) system that forces more air into the cylinder so that more fuel can be burnt in the same volume without increasing the AFR. The turbine is placed in the exhaust stream and uses the energy of the exhaust gas to drive the compressor connected on the same shaft. The compressor placed in the intake system compresses the air to be fed to the cylinders.

Another disadvantage of Diesel engine is its typical noise generated by the combustion. It is due to the steep pressure gradient immediately following the ignition delay period where a significant quantity of premixed charge ignites. This can be limited by injecting a small amount of fuel prior to the main injection. This so-called pilot injection has become a viable solution in recent modern injection systems. Alternatively, injection rate modulation can be applied with a modern CR injection system to achieve lower combustion noise.

Regarding emission matter, the overall CO, CO_2, and HC pollutant emissions of diesel engines are considerably lower than SI engines. On the contrary, NO_x emission is present in larger quantities in the tailpipe of a diesel engine and, due to its lean operations, its abatement cannot be operated through a TWC system (which requires stoichiometric air–fuel mixtures) as done in a SI engine. Therefore, NO_x traps and Selective Catalytic Reduction (SCR) catalysts have been introduced in order to decrease the NO_x emissions. Another reason of concern for this kind of engine is the Particulate Matter (PM) emission for whose abatement proper Diesel Particulate Filters (DPF) or traps are currently used on modern vehicles. Nevertheless, although in the last decades both NO_x and PM emissions have been drastically reduced, they remain at relatively high levels.

1.2.4 HCCI Engines

The Homogeneous Charge Compression Ignition (HCCI) is a combustion process rather than an engine type. It combines the features of the SI and CI engines [13]: the mixture of air and fuel is obtained by means of PFI or very early DI in order to have time to reach an homogeneous charge before starting the combustion stroke, as in SI engines; during the compression the pressure and thus temperature is increased until autoignition occurs, as in CI engines.

The first studies were presented in 1979 [14, 15], achieving autoignition in 2-stroke gasoline engines. 4 years later, a study of a Compression-Ignited Homogeneous Charge (CIHC) combustion process on a 4-stroke engine was presented in [16] where autoignition of the mixtures of fuel, air, and exhaust products charge was obtained in a single cylinder engine with variable compression ratio. Only in 1989, this new combustion process was called for the first time HCCI in [17].

The HCCI engine runs unthrottled, reducing the pumping losses like the CI engine. The combustion process starts when the premixed charge is autoignited during the compression stroke, due to high pressure and temperature reached in the combustion chamber. The charge is ignited in several locations, where the conditions are favorable for chemical reactions to occur, resulting in a fast heat release compared to SI and CI engines. Therefore, in order to keep the pressure rise rates at acceptable levels, the mixture is diluted by means of air or Exhaust Gas Recirculation (EGR). This leads also to decrease the combustion temperature reducing therefore NO_x and particulate matter but, on the other side, sometimes the fuel is not entirely oxidized causing increased emissions of CO.

Controlling the combustion phasing in an HCCI engine is an issue since it depends on temperature, pressure, and charge composition during the compression. In this regard, a monitoring of the combustion process is necessary through a proper sensing. To this aim, the use of in-cylinder pressure sensor is expensive for commercial vehicle and therefore limited to laboratory experiments. Alternatively, cheaper systems exploiting ionic current sensors have been recently proposed [18, 19]. Once the information on HCCI combustion is available, different methods for regulating the ignition timing are proposed controlling the intake air temperature or injecting different mixture fractions of two fuels with different ignition properties.

1.3 Engine Management System

The EMS is the software, running on the Electronic Control Unit (ECU), responsible for optimizing the performance and efficiency of the engine. The ECM collects all sensor data, interprets and processes this data, and then commands actuators necessary for the smooth and efficient operation of the engine, controlling the amount of fuel being injected and adjusting the ignition timing.

Before ECUs, engine operations were regulated by mechanical and pneumatic devices. An early example of a system using these devices for the automated and simultaneous management of several engine functions was designed by BMW in 1939 for their 801 14-cylinder aviation radial engine [20].

In the early 1970s, the engine management systems was a simple injection system with a separate ignition unit, evolving later during the 1980s when injection and ignition were integrated into one single ECU. Then, with the introduction of electronic throttle control as a drive-by-wire system, the EMS reaches the ability to control the generation of torque over the entire operating range of the engine.

Nowadays, a modern EMS is devoted not only to engine basic functions, such as injection and ignition timing or emission control (i.e., AFR closed-loop control or catalyst heating), but also manages additional systems such as continuous camshaft, resonance flap actuation or engine fan. A modern EMS must also be equipped with a complete onboard diagnostic and monitoring system communicating with transmission control units by means of the Controller Area Network (CAN) bus.

The new functional architecture of the EMS is characterized by the following main features [21]:

- *Centrally coordinated torque management*: all torque demands coming from internal functions or external systems (i.e., drive train or vehicle dynamics control) are collected by the EMS resulting in a variation of torque or efficiency.
- *Centrally coordinated A/F management*: similarly, all mixture demands are coordinated by one central manager.

Typically, an EMS contains all functions to control a modern SI engine. In the following, a brief functional overview is given considering only the main system features.

- The *engine torque management* controls all torque-influencing actuators. In the case of several torque requests, derived simultaneously from different subsystems, the functional structure is characterized by two coordination steps. The first step is to associate a priority to each torque request. The requests can be internal, for example generated by the start function, idle speed control, engine speed limitation, as well as engine protection functions or catalyst heating, or external, for example defined by the driver, cruise control, or vehicle dynamics control. According to the selected priority, in the next step the resulting torque demands are processed in driveability functions, dashpot function (to limit the minimal intake manifold pressure) and antijerking function. The driveability function is characterized by customer requests, varying between a comfortable and a sportive characteristic.
- *AFR control with a central A/F manager*. To achieve high efficiency of the automotive exhaust gas after-treatment device and so reduce tail pipe emissions, most of today SI-ICEs operates at stoichiometric AFR. Therefore, the main goal of the engine control system devoted to emission reduction is to regulate the AFR. Typically, the control strategy designed to this aim is composed of two actions.

A closed-loop control scheme obtained feeding back the oxygen content on the exhaust gas through a Universal Exhaust Gas Oxygen (UEGO) sensor. This measurement is affected both by uncertainty, caused by the kind of sensor mounted on the manifold, and by the delay on the signal, being the sensor placed on the exhaust manifold far from the valves to prevent its damaging due to high temperature reached by the gas close to the cylinders. The second action is realized through a feedforward controller, mainly based on the prediction of air incoming into the cylinders. Since the air mass flow sensor is a sensitive device that must be placed far from the intake valves, it cannot be used to measure directly the air charging the cylinders. Therefore, it is necessary to dispose reliable models able to compute the air charge based on indirect measurements, such as intake manifold pressure, engine speed, and air mass flow crossing the throttle valve.

- *Fuel injection*. In order to meet the severe international regulations on admissible pollutant emissions, the exhaust manifold has been equipped with the TWC converter that is able to reduce the quantity of CO, HC, and NO_x pollutants yielded during combustion process. The efficiency of this system is very high if the engine operates close to stoichiometric mixture ratio. Therefore, the open-loop control system used initially to supply the fuel during the combustion phase has been substitute with a more precise closed-loop regulator, capable of maintaining the air–fuel mixture close to stoichiometric condition both during steady-state and transient engine operations.
- *Ignition timing/angle*. The ignition angle is the crank angle where combustion starts and it has to be controlled in order to obtain a combustion free from knock. To this aim, special equipments are necessary to detect accurately the occurrence of knock phenomenon. A very sophisticated open-loop system, using many sensors, can be designed resulting in an elaborate map containing a very large numbers of ignition angles. Unfortunately, this solution is impracticable on commercial vehicles and it is substituted with a simpler open-loop system and a simpler map, providing ignition angles able to reach good performance and, at a same time, a reasonable margin of safety for the engine to keep it free from knock and damages. Generally, a feedback action can be added, generated by the knock sensor (i.e., an accelerometer sensor), allowing the adjustment of the spark timing: at the beginning, the timing is calculated according to a precalibrated map; the ECU then advances this basic timing setting, reducing it in small steps of crankshaft degrees (e.g., 1.5°), until accelerometer sensor indicates knock-free combustion is re-established. The accelerometer, usually mounted on the engine block, detects high-frequency vibrations (in the range of 5–20 kHz) propagating in the engine from those cylinders where knock phenomenon is originated.
- *Idle speed control*. Automotive idle speed control is a critical issue in engine control fields. Essentially it is a highly nonlinear and time-varying problem. Its performance has a significant impact on fuel economy and emission levels. Idling is one of the most often used functionalities in the modern car. This is especially the case in city traffic, where there are frequent stop and go situations. Therefore, improvements of the control performance for the idle speed control unit has always been a high priority. That is, keep the engine speed at a desired setpoint

value, ensure good disturbance rejection while maintaining low fuel consumption. Typical disturbances that are to be rejected by the controller are loads from the air-conditioning system or power steering. Obviously, the ECU compensates such disturbances on engine torque by using the throttle, however, due to the slow dynamics of the air mass in the intake manifold, this would generate an unacceptably slow disturbance rejection. For this reason, the spark advance is used as a second control variable, by advancing or retarding the ignition and obtaining an instantaneous torque variation from the engine. However, a deviation from the optimal spark ignition will result in higher fuel consumption. Thus, the use of this signal should be kept at minimum and used only for improving the speed of the disturbance rejection. From control point-of-view, this is a difficult problem since the system in question is nonlinear, multivariable (two inputs) and time varying. Moreover, the throttle control channel has a slower dynamics than the spark advance event. In the literature this problem is usually approached by treating the two control channels separately (one control signal is set to constant while the other is modified), leading to performance degradation. Some other approaches treat linearized models resulting in local designs. There exist approaches where both control signals are treated in the same time (multivariable control), however, the resulting controllers are highly complex and difficult to tune.

References

1. J.B. Heywood, *Internal Combustion Engine Fundamentals* (McGraw-Hill, New York, 1988)
2. C. Hirsch, *Numerical Computation of Internal and External Flows—Fundamentals of Numerical Discretization*. Wiley Series in Numerical Methods in Engineering, vol. 1. (Wiley, New York, (1997). ISBN 0-471-92385-0
3. C. Hirsch, *Numerical Computation of Internal and External Flows—Computational Methods for Inviscid and Viscous Flows*. Wiley Series in Numerical Methods in Engineering, vol. 2 (Wiley, New York, 1998). ISBN 0-471-92452-0
4. D.E. Winterbone, R.J. Pearson, *Theory of Engine Manifold Design: Wave Action Methods for IC Engines* (Wiley, New York, 2005). ISBN 978-1860582097
5. C. Stan, *Direct Injection Systems for Spark-Ignition and Compression-Ignition Engines* (SAE International, Warrendale, 1999)
6. F. Zhao, D.L. Harrington, M.C.D. Lai, *Automotive Gasoline Direct-Injection Engines* (SAE International, Warrendale, 2002). ISBN 978-0768008821
7. J.W. Oh, C.W. Park, H.S. Kim, G.B. Cho, Effect of multiple injection on the performance and emission characteristics of lean burn gasoline direct injection engines. Trans. Korean Soc. Mech. Eng. B **36**(2), 137–143 (2012)
8. M. Fry, J. King, C. White, A comparison of gasoline direct injection systems and discussion of development techniques. SAE Technical Paper (no. 1999-01-0171) (1999). doi:10.4271/1999-01-0171
9. G. Stumpp, M.J. Ricco, Common rail—an attractive fuel injection system for passenger car DI diesel engines. SAE Technical Paper (no. 960870) (1996). doi:10.4271/960870
10. O. Chiavola, P. Giulianelli, Modeling and simulation of common rail systems. SAE Technical Paper (no. 2001-01-3183) (2001). doi:10.4271/2001-01-3183

11. H. Tomishima, T. Matsumoto, M. Oki, K. Nagata, The advanced diesel common rail system for achieving a good balance between ecology and economy. SAE Technical Paper (no. 2008-28-0017) (2008). doi:10.4271/2008-28-0017
12. W.W. Pulkrabek, *Engineering Fundamentals of the Internal Combustion Engine*, 2nd edn. (Prentice Hall, Upper Saddle River, 2004). ISBN 0131405705
13. K. Epping, S. Aceves, R. Bechtold, J. Dec, The potential of HCCI combustion for high efficiency and low emissions. SAE Technical Paper (2002-01-1923) (2002). doi:10.4271/2002-01-1923
14. S. Onishi, S.H. Jo, K. Shoda, P.D. Jo, S. Katof, Active thermo-atmosphere combustion (ATAC): a new combustion process for internal combustion engines. SAE Technical Paper (no. 790501) (1979). doi:10.4271/790501
15. M. Noguchi, Y. Tanaka, T. Tanaka, Y. Takeuchi, A study on gasoline engine combustion by observation of intermediate reactive products during combustion. SAE Technical Paper (no. 790840) (1979). doi:10.4271/790840
16. P. Najt, D. Fosterì, Compression-ignited homogeneous charge combustion. SAE Technical Paper (no. 830264) (1983). doi:10.4271/830264
17. R.H. Thring, Homogeneous-charge compression-ignition (HCCI) engines. SAE Technical Paper (no. 892068) (1989). doi:10.4271/892068
18. P. Mehresha, J. Soudera, D. Flowersb, U. Riedelc, R.W. Dibblea, Combustion timing in HCCI engines determined by ion-sensor: experimental and kinetic modeling. Proc. Combust. Inst. **30**(2), 2701–2709 (2005). doi:10.1016/j.proci.2004.08.135
19. Y. Huang, D. Mehta, Investigation of an in-cylinder ion sensing assisted HCCI control strategy. SAE Technical Paper (no. 2005-01-0068) (2005). doi:10.4271/2005-01-0068
20. B. Gunston, *World Encyclopedia of Aero Engines* (Patrick Stephens, Cambridge, 1989). ISBN 1-85260-163-9
21. J. Gerhardt, H. Hönninger, H. Bischof, A new approach to functional and software structure for engine management systems—BOSCH ME7. SAE Technical Paper (no. 980801) (1998). doi:10.4271/980801

Chapter 2
Basic Concepts on GDI Systems

Abstract Gasoline Direct Engines offer many advantages as compared to PFI engines, as regard efficiency and specific power. To fully exploit this potential a particular attention must be paid to the in-cylinder formation process of air/fuel mixture. More demanding performance is required to the combustion system, since injectors must provide a fine fuel atomization in considerably short time, achieving a spray pattern able to interact with in-cylinder air motion and piston top surface. This is made possible through the Common Rail technology allowing an injection pressure one order of magnitude higher as compared with that of conventional PFI engines. Fuel economy can be obtained regulating load by mixture leaning, minimizing throttle usage at low loads where pumping losses are more significant, and requiring charge stratification for a stable ignition and combustion. Charge stratification can be pursued based mainly on the sole action of the fuel spray or on its interaction with a specially shaped surface on piston top or with the air bulk motion. Depending on the modality of stratification attainment, different combustion systems can be considered. The injector design has in turn a key role being the final element of fuel metering required to the desired spray pattern, injected fuel mass per injection event, resistance to thermal stress and deposits. Injector housing and orientation with respect to the combustion chamber has to be carefully chosen, exploiting in this regard the indications of computational fluid dynamics (CFD), provided by 3D simulations. Some fundamental scheme is provided for the whole high pressure fuel delivery plant, as employed in current vehicles equipped with GDI spark ignition engines.

G. Fiengo et al., *Common Rail System for GDI Engines*, SpringerBriefs in Control, Automation and Robotics, DOI: 10.1007/978-1-4471-4468-7_2, © The Author(s) 2013

2.1 Mixture Preparation: Gasoline Port Fuel Injection Versus Direct Injection

A short comparison between gasoline Port Fuel Injection (PFI) and direct injection (DI) can be made to highlight their differences, advantages, and drawbacks as regard mixture preparation and its consequences on performances, fuel economy, and emission. In PFI engines with an injector for each cylinder, the injection system is quite simple and without demanding requirements. Fuel is fed to the injectors by a common manifold at a pressure usually in the range of 3–5 bar, maintained fairly constant thanks to a pneumatic regulator driven by the absolute pressure in the inlet manifold. Depending on injector features, such a pressure allows fuel atomization with a droplet Sauter mean diameter (SMD) in the range of 120–200 micrometer. The SMD is defined as the diameter of the droplet having the same surface to volume ratio as that of the overall spray. Assuming x as the diameter of a single droplet, and being $f(x)$ the corresponding density probability function, SMD is defined as

$$\mathrm{SMD} = \frac{\int_{x_m}^{x_M} x^3 f(x)\mathrm{d}x}{\int_{x_m}^{x_M} x^2 f(x)\mathrm{d}x},$$

where x_M and x_m are the maximum and the minimum droplet diameter, respectively. Fuel delivery is performed out of the cylinder, with the scope of obtaining a fairly homogeneous charge, and the spray produced by the injector is directed onto the inner wall of the inlet duct and the back of the inlet valve, forming a fuel puddle, and mixture preparation occurs by fuel evaporation from the puddle and the suspended droplets. Even if the engine is fully warmed up, depending on the amount of injected fuel, it is unlikely that the time interval prior inlet valve opening is enough to complete mixture preparation out of the cylinder. Especially if temperatures of duct wall and inlet valve and available time are not adequate for complete external evaporation (for example at cranking and cold start), a portion of fuel is inducted into the cylinder as liquid rivulet and droplets, while some residual remains on the inner surface of the inlet duct. This means that the fuel quantity inducted in the cylinder is not that injected by the injector, and a delay occurs in fuel delivery. Several consequences follow: imprecise fuel metering; time response worsening in transient maneuvers; incomplete combustion and soot formation due to inlet valve seat wetting and incomplete in-cylinder droplet evaporation. In this regard, cold cranking is a critical condition where low temperature and low turbulence hinder evaporation; it may be necessary to deliver a fuel amount up to 4–5 times the stoichiometric one, with delays in engine start and high level of engine out emission as unburned hydrocarbons (UHC) and carbon monoxide (CO). In gasoline DI (GDI) engines air only is inducted through the inlet valve, and mixture is prepared delivering fuel in the cylinder. As already pointed out, this solution is able to promote charge cooling thanks to heat absorption by fuel evaporation, leading to a better volumetric efficiency and a lower propensity to knock and thus to the possibility of higher compression ratio and higher efficiency. Furthermore, a more precise fuel metering is obtainable, giving the possibility to operate with lean mixture and to control load by means of mixture strength variation, avoiding the pumping losses due to throttling at low loads, with beneficial effect on efficiency. On the other hand, there is the need to complete mixture preparation in the time interval between inlet valve opening and ignition triggering by spark, which is much lower than the analogous time allowed in PFI systems. For this reason, more demanding requirements have to be satisfied by (GDI) systems. Actually, fuel evaporation in so a short time requires a fine atomization, featured by a droplet SMD not higher than 20 μm. This is the reason why specifically designed injectors are needed with an injection pressure one order of magnitude higher than that of PFI systems. This is accomplished in current (GDI) engines using common rail (CR) technology which is able to provide injection pressures well beyond 100 bar. However, despite the finer atomization, the possibility of surfaces wetting (e.g. cylinder wall and piston crown) cannot be avoided; actually sometimes it is wanted in conjunction with some interactions with the in-cylinder air flow to promote charge stratification. The possibility of more precise fuel metering offers to (GDI) engines the advantage of low mixture enrichment at cold cranking, with consequent lower UHC emission and a more rapid starting. Basically, depending on the working condition, air/fuel mixture in (GDI) engines can be homogeneous (that is with an fairly uniform AFR in the combustion chamber volume) or stratified (that is with AFR space gradients, richer in the volume near the spark plug and progressively leaner as distance from spark plug increases), a result attained acting on injection timing. Homogeneous charge can be obtained by an early injection, actuated at the beginning of the air induction phase, so as to allow an adequate evaporation and mixing of fuel in conjunction with the air motion promoted by the inlet ducts geometries. Generally such a condition is promoted at high load operation. Charge stratification can be attained by late injection, actuated during the upward piston stroke of compression, metering the fuel to obtain an overall lean AFR ratio. Charge leaning, if not too high, allows load control without throttling, with beneficial effect on fuel economy. Stratification is motivated by the need to have a richer mixture near the spark plug gap for a stable ignition, otherwise misfiring could occur.

2.2 GDI Combustion Systems

The attainment of a desired stratified charge have to take into account the reciprocal position of spark gap and injector. Generally, a central position of the spark plug in the combustion chamber is preferred: this choice is motivated by the need to reduce the probability of knocking combustion, occurring when the unburned mixture furthest from spark gap reaches auto-ignition before the arrival of the flame front. A central position of the spark plug allows a symmetrical propagation of the flame front initiated by the spark, performing a shorter path to extend combustion to the whole unburned mixture before auto-ignition occurs. Twin spark engines are equipped with a second spark plug in peripheral position, giving a further support to enhance knock prevention (see also Fig. 2.2).

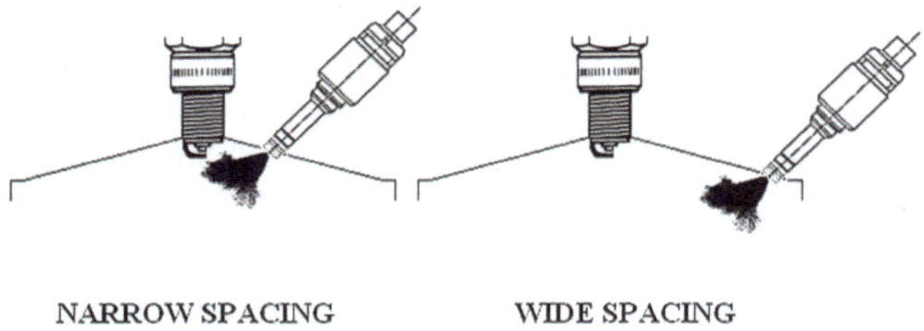

Fig. 2.1 Scheme of the different locations of injector

Fig. 2.2 Example of injector location between intake valves, feft and central holes are for the housing of spark plugs (Twin-Spark engine)

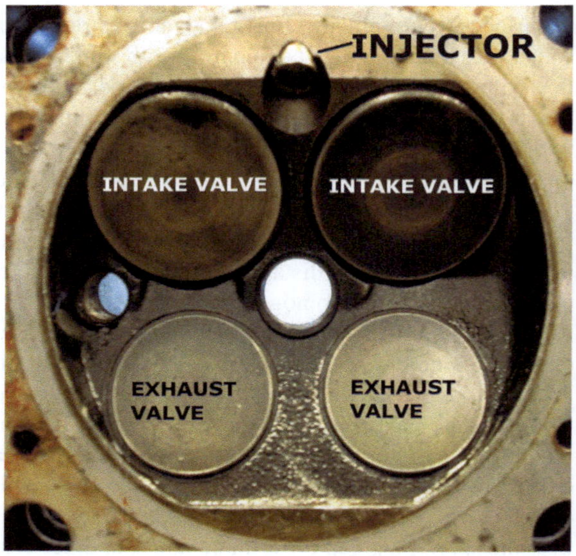

Injector can be located in proximity of spark gap (narrow spacing, see Fig. 2.1); this choice is more favorable for stratification attainment since this result can be pursued more easily without involving the motion of the air charge bulk. The main disadvantage of this solution is the higher probability of spark plug fouling caused by electrodes fuel wetting; furthermore, fuel injector location in central position or with a small offset has to be carefully chosen considering the higher thermal stresses due to exhaust valves vicinity. As alternative to the previous solution, fuel injector can be located away from the spark gap (wide spacing, see Fig. 2.1); in this case the sprayed fuel plume has to perform a longer path to reach the gap. This results in a longer time interval for stratification attainment and in the requirement of proper interaction with the air charge bulk and the moving surface of the piston. Generally, in a four valves engine head the position is chosen between the inlet valves, where injector tip overheating is avoided and a more effective interaction is possible between the fuel spray and the air flow incoming from inlet ducts (Fig. 2.2).

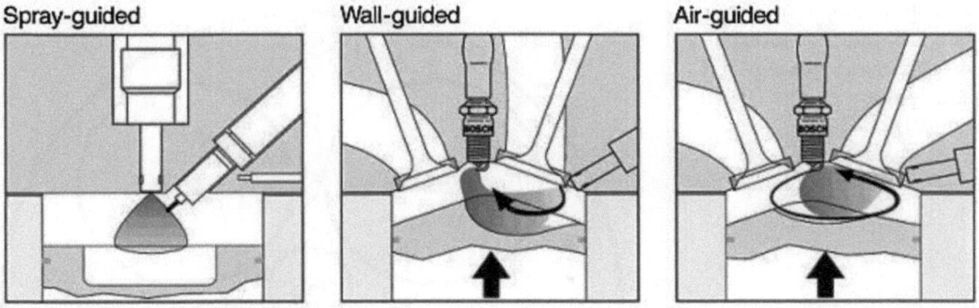

Fig. 2.3 Scheme of the different combustion systems (taken from [1])

In this regard, the correct matching between injector features and in-cylinder air motion is fundamental to attain the desired stratification, with the AFR required for ignition in the volume around the spark plug gap, and at the time of spark generation. To comply with these demanding requirements based on experiments only would be an expensive and time-consuming task; therefore, computational fluid dynamics (CFD) simulations, both 1D and 3D, are a valuable aid to search more quickly the optimal parameters for the design of combustion chamber, inlet and exhaust ducts, and injector. Of course, this does not eliminate the need of final experimental validation of the results produced by simulation, usually referred to specific operating conditions. This interaction between simulation and experiments is useful to setup the best-compromise solutions for the whole operating range of an engine, certainly allowing a significant reduction of development costs and time to market.

The combustion systems can be distinguished depending on the way used to obtain charge stratification. Thus, the following classification is possible (Fig. 2.3):

- Spray guided systems, where the stratification process relies mainly on the features of the spray and its dynamics, as regard fuel atomization, droplet distribution, and overall geometry;
- Wall guided systems, where the stratification process is based mostly on the interaction between the fuel spray and the surface of a proper cavity on the top of piston;
- Air guided systems, where stratification is obtained mainly by means of the interaction between the fuel spray and the motion of the air charge inducted in the cylinder.

Actually, a real engine can be classified according to the features of one of the previous system, but this does not exclude the possibility that some of the features of the other ones can be recognized.

In the following, a brief description of the main features of these system will be provided.

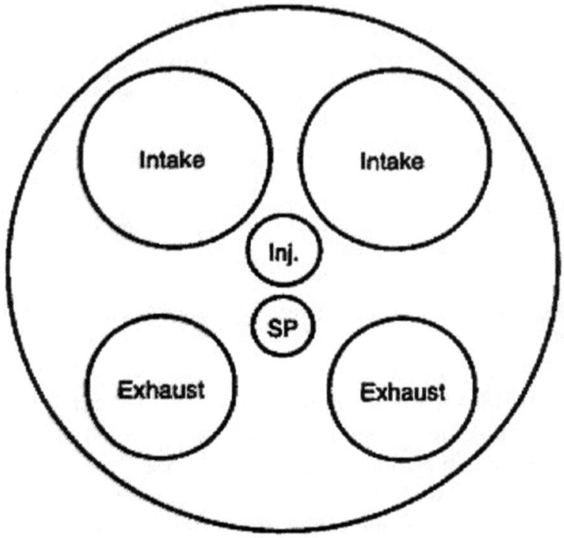

Fig. 2.4 Example of injector and spark plug location in a 4 valve spray guided GDI spark ignition engine (taken from [2])

2.2.1 Spray Guided Systems

As previously outlined, in spray guided systems charge stratification results mainly from the spray dynamics, with minor contribution from the interaction with the bulk air charge motion and the surface of the piston cavity. To this aim, injector has to be located in a position close to the center of the cylinder head and near the spark plug (Fig. 2.4), giving the advantage of a symmetrical distribution of fuel in the combustion chamber with a better utilization of air charge.

There are several drawbacks of this system; one of them is the higher tendency to spark plug fouling and soot formation, as well as the sensitivity to spray geometry variations, as consequence of off-design operation of injectors resulting, for example, from deposits or CR pressure deviations with respect to the value required for the desired performance. These spray variations can lead to higher Indicated Mean Effective Pressure (IMEP) coefficient of variation and misfiring. In relation to the central position of the injector, some problems arise for unwanted fuel impingement and consequent wetting of piston head, occurring with late injection when piston approaches top dead center in the upward stroke. A way to mitigate this drawback is a combustion chamber design providing a bowl in the piston top (Fig. 2.5) or a dome in the cylinder head (Fig. 2.6), so as to increase the path of the fuel plume toward the piston surface.

Another issue in spray guided systems, again related to spark plug proximity, is represented by housing difficulties in the cylinder head, forcing designers to reduce the diameter of inlet valves, with consequent penalization of air charge intake. The problems related to the sensitivity to spray variation can be relieved by the adoption of air assisted injectors; in this case, with the injector aligned with cylinder axis, fuel distribution is obtained by pneumatic atomization. This solution reduces

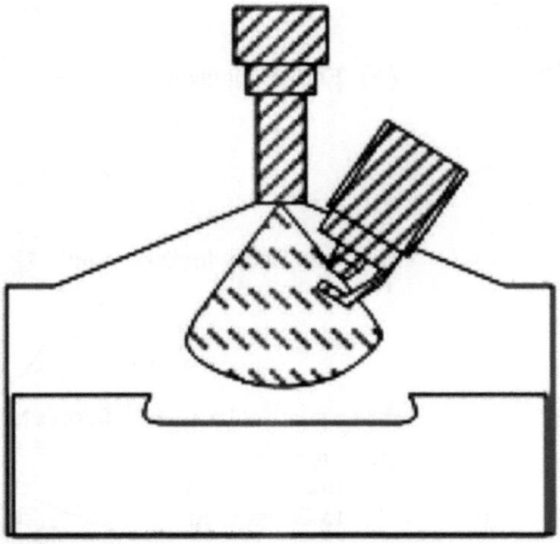

Fig. 2.5 Example of injector and spark plug location in a 4 valve spray guided GDI spark ignition engine (taken from [3])

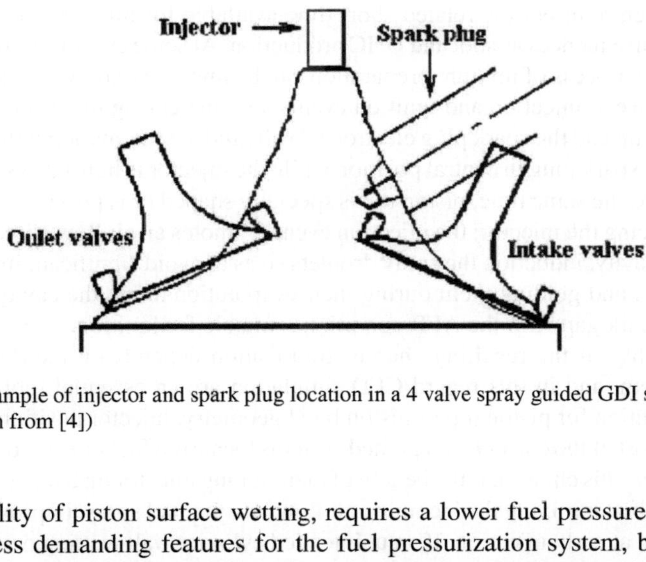

Fig. 2.6 Example of injector and spark plug location in a 4 valve spray guided GDI spark ignition engine (taken from [4])

the possibility of piston surface wetting, requires a lower fuel pressure difference, allowing less demanding features for the fuel pressurization system, but involves other complications for the presence of an air compressor and a storage plenum.

2.2.2 Wall Guided Systems

A stable charge stratification can be pursued maintaining a short distance between injector tip and spark gap, as well as a short time interval from injection start

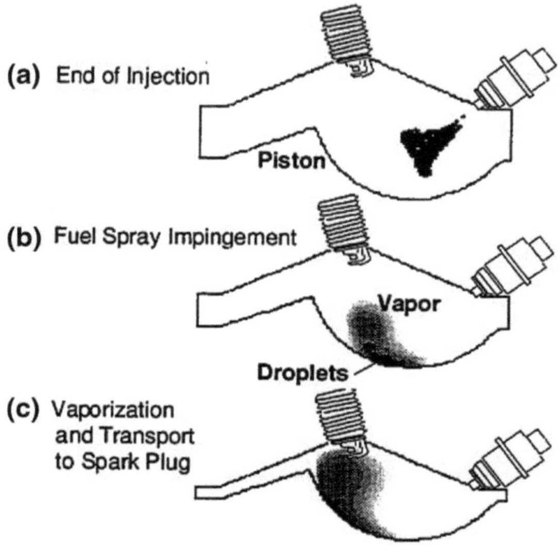

Fig. 2.7 Scheme of the operating phases of a wall guided combustion system (taken from [5]).

and spark ignition; but the related short time available for mixture preparation has negative consequences as soot and UHC production. At some cost on stratified charge stability, the process of mixture preparation can be improved allowing a greater time interval between injection and ignition events, and increasing the distance between the injector tip and the spark plug electrode. Wall guided systems adopt this approach locating the spark plug in central position while the injector is housed in side position (Fig. 2.7). At the same time, piston top is specially shaped or is provided with a bowl or cavity facing the injector: the injection event promotes an air flow, directed toward the piston cavity, induction the spray droplets so as to avoid significant impingement and wetting, and guiding them during their evaporation along the cavity surface to reach the spark gap with the AFR conditions suitable for ignition.

The quality of the resulting charge stratification depends on the design of the whole system, and in this regard CFD simulation are an essential tool to find the optimal solution for piston top or piston bowl geometry, injector position, and spray features. Most of the current wall guided systems locate the fuel injector on the side of intake valves; this choice has the benefit of maintaining injector tip temperature at safe level, and a better mixing the incoming air flow into the fuel spray, improving droplet evaporation, at early injection. If an unintended cylinder wall wetting occurs at early injection, an increase in UHC emission follows because of incomplete evaporation and mixing with air and of adsorption and subsequent desorption of the fuel that, after being trapped in the piston top land and inter-ring crevices, is dissolved in oil with consequent dilution and lose of lubricant properties. Such a condition occurs for example, at high load and speed when CR mean pressure is increased to meter a higher fuel amount, and possible pressure pulsations in the CR systems make the injector operate with spray penetration exceeding the design limits. This underline

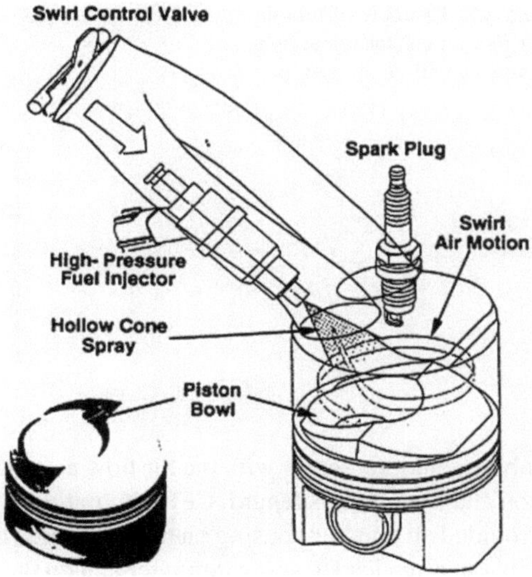

Fig. 2.8 Example of swirl air motion coupled with a bowl in the piston (taken from [6])

the importance of a control able to regulate CR pressure in all the possible engine operating conditions. The design of piston bowl is crucial to attain the desired charge stratification, however, in all cases a compromise solution has to be adopted. For example, the choice of a deep bowl allows to extend the operating range where engine is able to work with stratified mixture, but this is counterbalanced by several drawbacks. First, with respect to a flat piston an actual homogeneous charge at wide open throttle is more difficult to be obtained with a deep bowl shaped piston, resulting in combustion worsening and reduction of full load torque. Second, the more complex piston shape, apart from higher manufacturing difficulties and costs, leads to higher heat losses from burned gas affecting negatively engine efficiency. In-cylinder air motion has a significant role in the guiding of the fuel cloud produced by the injector along the piston cavity walls toward the spark gap, as well as for the evaporation of the fuel droplets impinged on cavity surfaces. Basically, in-cylinder air motion can be of swirl type, featured by a rotation around an axis parallel to that of cylinder, or of tumble type, where the air flows rotate around an axis perpendicular to that of cylinder.

Swirl air motion is chosen when piston top design provides a bowl or an open chamber (Fig. 2.8). Reverse tumble motion, where the air flow rotates from the bottom to the top, is generally combined with a piston cavity shaped to receive the fuel plume during the upward piston stroke, flowing air and redirecting the resulting mixture toward the spark plug gap housed in central position (Fig. 2.9). The piston top side opposite to the cavity can be shaped in turn to promote a squish flow at the end of compression stroke, enhancing the tumble motion after ignition with effect on flame front propagation. Injector location and axis inclination is another crucial parameter to be considered, requiring a careful analysis of the whole injection process and its

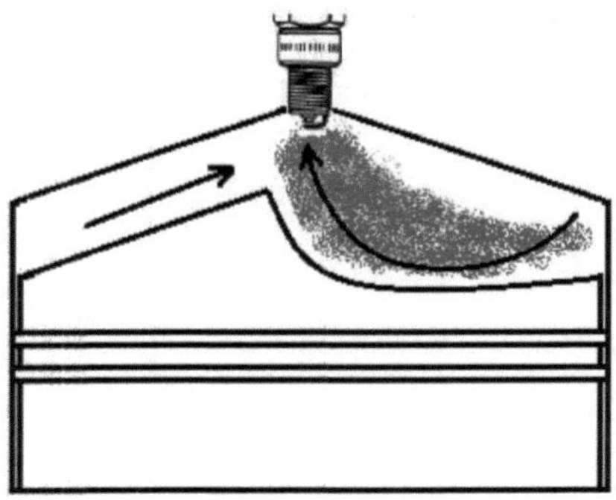

Fig. 2.9 Example of tumble air motion enhancement by a shaped cavity in the piston

subsequent interaction with the air flow and the geometrical features of the combustion chamber. In this regard, CFD simulation is a fundamental and indispensable aid. Provided an injector housing on the intake valve side, a low inclination of its injection axis can often lead to spray impingement on the opposite cylinder wall; further on, the outer region of the spray can impact against the intake valves during early injection. Predictable consequences are fuel deposit at piston top land crevices and valve seats, leading to UHC emission increase and soot production. As countermeasures there can be chosen an offset spray or a higher injector axis inclination; the latter solution has the collateral benefit of a good interaction between the spray and the piston cavity independently on piston position, with the possibility of stratified charge operation in a wider speed range. However, as the injector axis inclination is increased towards the vertical position, a significant worsening of mixture homogeneity can occur with penalization of full load torque and specific consumption. Special attention has to be paid to house injector tip neither projected in the combustion chamber, with consequent risk of intake valve wetting and deposits formation, nor in a recess though small, which could collect fuel droplets at spray periphery again with wetting and carbon deposit formation, but flush with the surface of combustion chamber.

2.2.3 Air Guided Systems

In air guided systems charge stratification is pursued based on the mutual action between the injected spray and the bulk air charge motion. Spark plug is still in central position while the injector is located in side position; air swirl or tumble motion is generated by means of the orientation and geometry of the inlet ports and ducts. Sometimes the desired motion can be enhanced by proper devices, like baffles, in the inlet ducts (Fig. 2.10).

In systems with two inlet valves, if allowed by the valve train mechanism, swirl could be induced by deactivation of one of them. Since charge stratification depends mainly on air motion in the cylinder and not on spray impingement, there is the

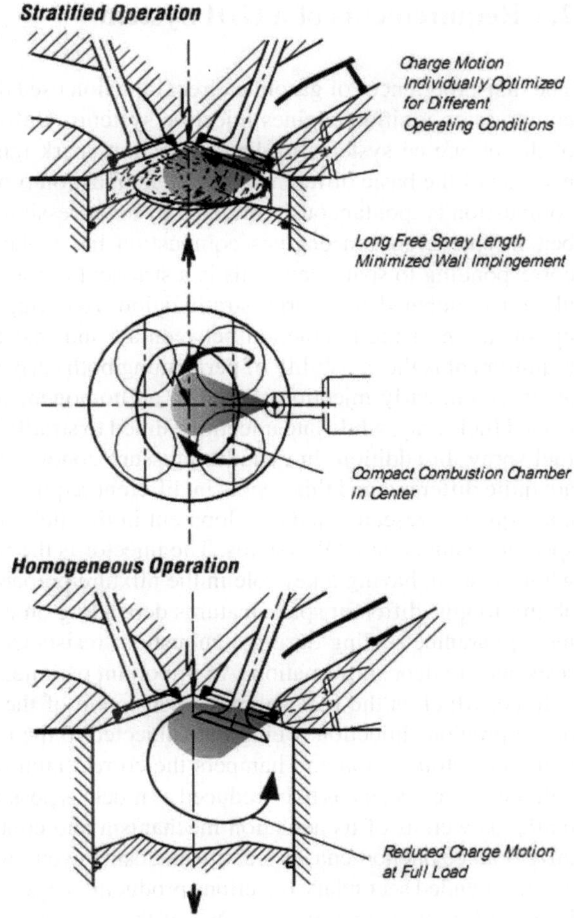

Fig. 2.10 Scheme of air guided system based on tumble motion. Notice the tumble control valve housed in the inlet duct (taken from [7])

advantage of avoiding UHC emissions related to fuel wetting of piston surfaces. This is counterbalanced by a higher sensitivity of combustion to changes of the air flow field, requiring a careful design of the combustion system considering the interaction between combustion chamber shape, inlet and exhaust geometry, and injector features. Despite in principle air guided systems do not require a special shape on piston top, actually some modification are made to ensure the attainment of the desired air motion, and this causes some difficulties in maintaining homogeneous charge at wide open throttle. As consequence, these systems too suffer of penalization of full load performance in terms of torque and specific fuel consumption.

2.3 Requirements of a GDI System

The first experiences of gasoline direct injection used devices basically derived from compression ignition engines injection systems. Unfortunately, a direct conversion of diesel engine systems to direct injection spark ignition engines is not possible because of the basic differences in the combustion process. While in diesel engines combustion is spontaneously initiated by compression in several zones of the chamber, in spark ignition engines combustion has to be triggered in a precise point corresponding to spark gap. This is a stringent constraint if unthrottled engine regulation is pursued by charge stratification, requiring both a spatial and temporal optimization of the interaction between air and injected fuel. Another demanding requirement is the capability of performing both early and late injection with proper atomization: early injection, aimed to an homogeneous charge, needs a more dispersed fuel spray, while late injection, aimed to stratification, requires a more narrow fuel spray. In addition, in-cylinder pressure conditions for early and late injection are quite different, and this results in different requirements for CR fuel pressure. As consequence, research and development in this field have been addressed to obtain specific features on GDI systems. The injector is the most important element of the whole system, having a key role in the mixture preparation process. Apart from the ability to give different spray features depending on engine operating conditions, it must guarantee sealing during combustion; resistance to temperature and leakage; resistance to deposit formation. An important parameter of injector design is the sac volume, which is the fuel volume downstream of the tip sealing element, residual of the previous injection. Being not subjected to the CR high pressure, it is ejected with poor atomization and hampers the correct atomization of the subsequent fuel volume; therefore, it has to be reduced as much as possible. The design of the injector pintle, as well as of its actuation mechanism and control, has to be aimed to minimize bounce phenomena, especially at closing event, which can lead to uncontrolled and unintended secondary injections producing larger droplets responsible of higher UHC and particulate emission. One of the basic parameters of GDI injector is the static flow capacity, corresponding to the volume flow rate resulting at pintle fully open and rated CR pressure condition. Based on this flow capacity, fuel is metered in the same way as in PFI systems, that is by injection duration which can range from 1 to 6 milliseconds. This is accomplished using a square wave voltage signal, defined as pulsewidth (PW) signal, generated by an ECU, and routed to an injector driver module. This module turns the PW signal into a current signal which is properly shaped to attain a fast opening and the desired hold-on condition. At a fixed CR pressure, a linear relationship exists between the duration of the input PW signal and the fuel amount metered per injection event, representing the injector characteristic. This linear function occurs over a range between a minimum and a maximum PW value, depending on injector design, and constitutes the operating range of the injector. By setting a different CR pressure, a different linear characteristic can be obtained; the higher the pressure the higher the characteristic slope as it will be shown later on (see Fig. 3.12a in Sect. 3.3).

By using a CR pressure regulator properly controlled by the EMS, it is possible therefore to have at disposal a wider dynamic range of the injector; for example, by setting a lower CR pressure it is possible to meter small fuel quantities with short injection duration using still a linear dependence between injected fuel mass and injection time. CR pressure level is set according to the desired features of the fuel spray, as penetration length, droplet SMD and distribution, cone angle, and so on. Fuel mass, ranging from some units to several tens of milligrams, is metered depending on the injection duration and on the difference between CR pressure and in-cylinder pressure. Any variation of this difference, which cannot be evaluated and taken into account in absence of an in-cylinder pressure measurement, leads unavoidably to metering errors. Further on, an unwanted change in CR pressure results in a transient change of the working characteristic of the injector, which in turn induces metering errors and alteration of the spray features from the desired ones. In this regard CR pressure pulsations, due to cyclic and sequential opening and closing of injectors as well as to high pressure pump operation, can contribute to combustion irregularities. The injection process is related to the fuel pressure in the injector, which in turn depends on CR pressure. The choice of its level is made combining the requirements of mixture formation, design features of the injector, parasitic load of high pressure pump, overall noise of the injection system. Often a compromise solution has to be searched. For example, a higher injection pressure on one hand can increase fuel atomization, on the other hand can lead to excessive spray penetration with unintended impingement on cylinder wall surface. On the contrary, for some type of injector, a higher injection pressure produces smaller droplets which spread and slow down too rapidly with reduction of spray penetration. CR pressure has a crucial role in cranking operation mode too; since the high pressure pump needs several crankshaft revolution to develop the steady rated pressure, during this transient condition the lower injection pressure of the tank fuel pump is available to meter the required fuel amount. This result in worsening of fuel atomization with an increase of droplet Sauter mean diameter up to 100 μm; injection duration increases as well and at cold start condition, when a higher fuel mass is required, it can exceed the maximum time allowable in relation to the intake phase. In order to comply with the requirement of a quick engine starting and low cold start emission, especially as regards UHC, injector must be designed to take count of these unfavorable conditions. As already mentioned, the injector has to provide a fuel atomization with a SMD not higher than 20 μm. Related to gasoline physical properties, this leads to the necessity of providing the CR with pressure levels up to 120 to 130 bar depending on the in-cylinder backpressure. Besides the effect on fuel atomization, the high pressure level, as compared to PFI systems, is motivated by the requirement of metering the needed fuel mass in a very short time interval at high speed and load. However, accurate fuel metering with a high injection rate is important too for stratified charge at low load: in those operating conditions the injector may work outside of the range of the linear relationship between injection duration and injected fuel mass, that is beneath the minimum value. In this case, when injection pulse duration is shorter than the minimum of the linear range, a PW lookup table is provided by a specific calibration. Another crucial injector parameter is the resistance to the formation

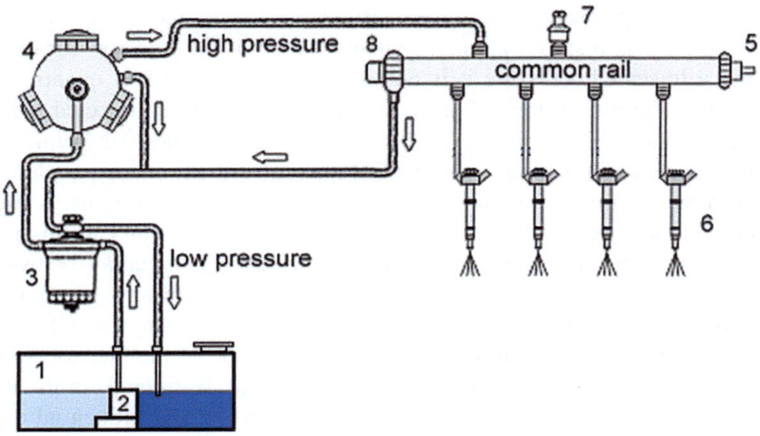

Fig. 2.11 Common rail injection system description: *1* fuel tank; *2* low pressure pump; *3* fuel filter; *4* high pressure pump; *5* common manifold; *6* electro-injector; *7* pressure sensor; *8* pressure regulation electro-valve

of deposits as related to drippage, that is the tendency to the accumulation of fuel resulting from defective closure or fuel droplets remained near the tip.

2.4 Description of the Common Rail Injection System

The main objective of a common rail system is to supply the electro-injectors with high pressure fuel independently by the amount to be injected. Therefore, the main benefit is that to decouple the regulation of the pump by the functioning of the injectors unlike the traditional injection systems where the mechanical pump generates a pressure that depends on the amount of fuel to spray.

A schematization of a common rail plant for spark ignition engines is shown in Fig. 2.11. The injection system is mainly composed by two separated sections: a low pressure circuit, consisting of a fuel tank, a fuel feed pump with a downstream filter, a low pressure pipe, and high pressure circuit formed by a high pressure pump, a high pressure line with a pressure sensor, a pressure regulator valve, a flow stopper, and the injectors. The low pressure electro-pump **(2)** forces the fuel from the tank **(1)** toward the high pressure mechanical pump **(4)** crossing the filter **(3)**, which cleans the fuel from impurity. The second pump compresses the fuel and sends it into the common manifold **(5)** (named common-rail) equipped with the electro-injectors **(6)**. The manifold is designed in order to hold low the pressure oscillations in the fuel due to the pump and the intermittent working of the injectors. Finally, the pressure in the manifold is regulated through the sensor **(7)** and the electro-valve **(8)** that flows the excess of fuel back into the tank.

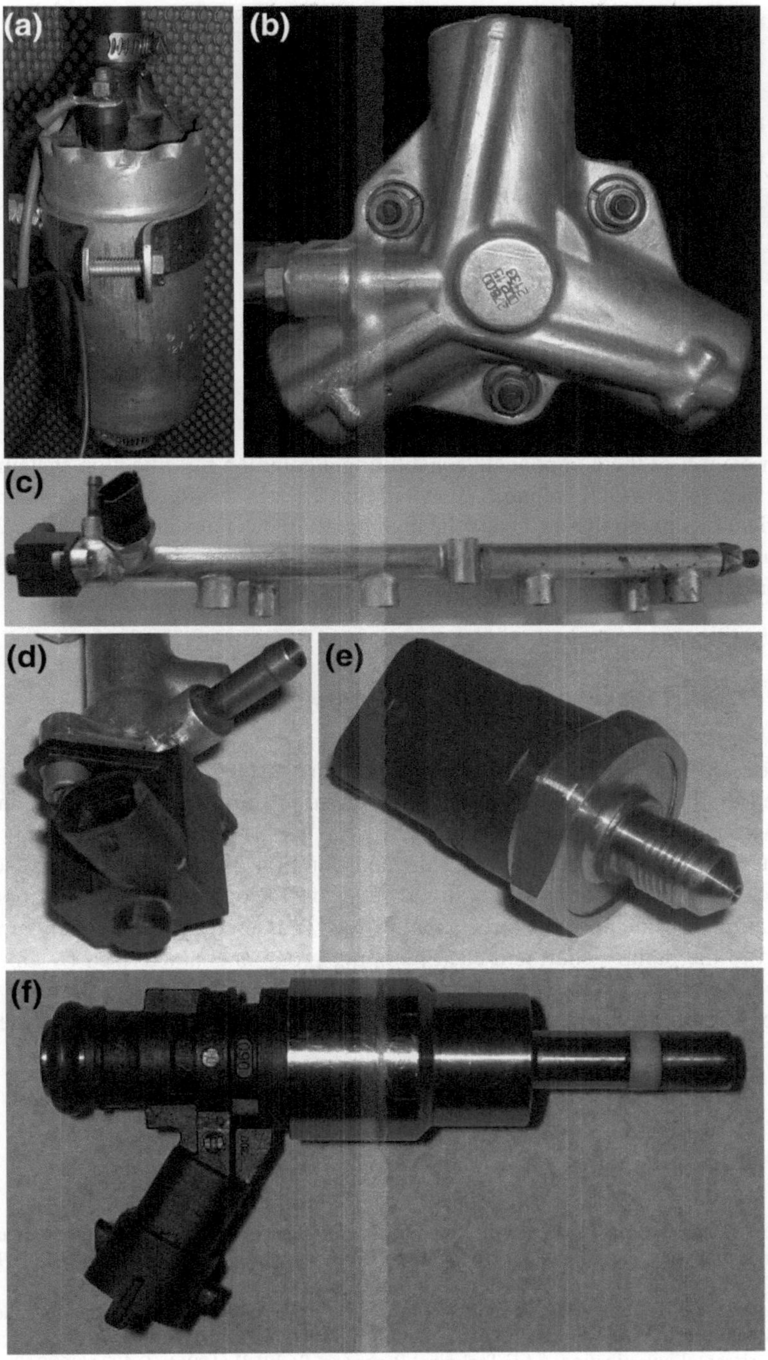

Fig. 2.12 Pictures of components common rail injection system mounted on a 2.0-liters GDI engine: **a** low pressure electrical pump; **b** high pressure mechanical pump; **c** common fuel rail; **d** electrovalve; **e** high pressure sensor; **f** electro-injector for direct injection of fuel

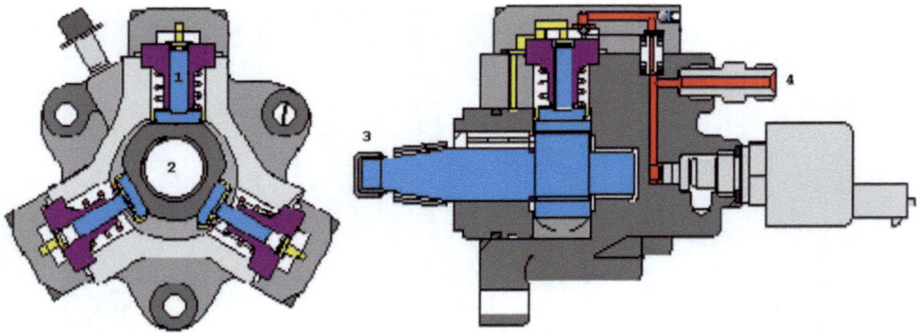

Fig. 2.13 High pressure pump scheme (taken from [8]): *1* piston; *2* triangular cam; *3* shaft; *4* exit hole

An overall of the components constituting a common rail injection system used to feed a 2.0-liters GDI engine is shown in Fig. 2.12. In detail, we have depicted: the electrical fuel pump in Fig. 2.12a; the high pressure mechanical pump in Fig. 2.12b; the fuel rail line in Fig. 2.12c; the control electro-valve in Fig. 2.12d; the pressure sensor in Fig. 2.12e; the solenoid injector in Fig. 2.12f.

2.4.1 High Pressure Mechanical Pump

Figure 2.13 shows a drawing of the high pressure mechanical pump. This pump is formed by three small pistons arranged in radial position (radial-jet) at an angular distance of 120 °. The pump is motored by the engine through the camshaft without the need of phasing since the start and duration of injection are imposed by the electronic control unit, which directly controls the opening of the injectors. The alternating movement of the three small pistons is assured by a triangular cam connected to the pump's shaft and each pumping group is characterized by an intake and exhaust valve. The combined action of the three pumping groups allows to reach high pressure values, up to 100–120 (bar), maintaining low level of residual pressure into the external manifolds.

References

1. C. Preussner, C. Dring, S. Fehler, S. Kampmann, Gdi: Interaction between mixture preparation, combustion system and injector performance. SAE Technical Paper (no. 980498) (1998).10.4271/980498
2. W. Anderson, J. Yang, D.D. Brehob, J.K. Vallance, R.M. Whiteaker, Understanding the thermodynamics of direct injection spark ignition (disi) combustion systems: An analytical and experimental investigation. SAE Technical Paper (no. 962018) (1996).10.4271/962018

3. M. Kawamoto, T. Honda, H. Katashiba, M. Sumida, N. Fukutomi, K. Kawajiri, A study of center and side injection in spray guided disi concept. SAE Technical Paper (no. 2005–01-0106) (2005).10.4271/2005-01-0106
4. T. Georjon, E. Bourguignon, T.D.B. Delhaye, P. Voisard, Characteristics of mixture formation and combustion in a spray-guided concept gasoline direct injection engine. SAE Technical Paper (no. 2000–01-0534) (2000).10.4271/2000-01-0534
5. Y. Iwamoto, K. Noma, O. Nakayama, T. Yamauchi, H. Ando, Development of gasoline direct injection engine. SAE Technical Paper (no. 970541) (1997).10.4271/970541
6. Y. Takagi, T. Itoh, S. Muranaka, A. Iiyama, Y. Iwakiri, T. Urushihara, K. Naitoh, Simultaneous attainment of low fuel consumption high output power and low exhaust emissions in direct injection si engines. SAE Technical Paper (no. 980149) (1998).10.4271/980149
7. J. Geiger, M. Grigo, O. Lang, P. Wolters, P. Hupperich, Direct injection gasoline engines—combustion and design. SAE Technical Paper (no. 1999–01-0170) (1999).10.4271/1999-01-0170
8. K. Ahlin, Modelling of pressure waves in the common rail diesel injection system. Ph.D. thesis, University of Linköping, Sweden, 2000

Chapter 3
A Control Oriented Model of a Common Rail System

Abstract A mean value model, particularly suited for control applications, of a common rail injection system for GDI engine is presented in this chapter. The mathematical model describes the electrical dynamics of the electrovalve, the actuation circuit, and the steady pressure of the fuel rail. Then, the mathematical model is validated through the comparisons with experimental data collected on a common rail system used to feed a 2.0-liter four cylinder GDI engine. Qualitative analysis of the effects that not manipulable inputs (disturbances) can have on the common rail pressure is furthermore investigated. Indeed, how battery voltage and engine speed variations can affect fuel pressure has been analyzed by means of parameter sensitivity functions (at steady conditions) and plant model simulations (at engine cranking), whereas the pressure ripple originated by the rotation of three lobes pump and functioning of the injectors have been characterized through a wide spectrum analysis of the alternating pressure component at different steady working conditions.

3.1 Mathematical Model

The CR system can be assumed as a quenched oscillating system in which the pressure waves generated by the first injection pulse produce a variation in the injection pressure of the subsequent injections [1, 2].

Many CR injection models proposed in the literature are based on the equations of the physics underlying the process or alternatively developed through simulation packages (see for instance [3–6]). A rather complete mathematical model for the numerical simulation of the common rail injection-system dynamics was developed in [7] to support experimentation, layout, and control design, as well as performance optimization. The good accuracy of all these models is counterpoised by the complexity of the resulting equations. Hence, those models are suitable for mechanical design and fluid dynamic prediction of the system behavior but they result too complex to be used for control application purposes.

G. Fiengo et al., *Common Rail System for GDI Engines*, SpringerBriefs in Control, Automation and Robotics, DOI: 10.1007/978-1-4471-4468-7_3, © The Author(s) 2013

Table 3.1 Estimated parameters of the electrovalve model

Parameter	Unit	Value
R	Ω	5.4
L	mH	18.7

Accurate nonlinear mathematical models of the common rail injection systems for diesel engines were developed in [8] and [9] in the same year. Both models were obtained by applying elementary fluid dynamics, mechanics, and electrical laws to the subcomponents which injection system was decomposed. Although these are simpler than one-dimensional (1D) models, they have never been used for designing model-based pressure controllers as for as we are aware. Later on, an innovative hybrid modeling approach in automotive control application was proposed in [10] to describe the discrete–continuous interactions of a common rail fuel injection system for multijet diesel engines (due to the slow time-varying frequency of the HP pump cycles and the fast sampling frequency of sensing and actuation). The model was then exploited to design a multirate hybrid control algorithm.

More recently, a very simple but effective mathematical model was developed by the authors [11] to describe the in-cycle average pressure of a common rail injection system for GDI engines. This model takes into account the electrical dynamics of the command electrovalve coupled with its actuation circuit, and the steady fuel pressure when engine speed and electrical current vary. The battery voltage level is furthermore taken into account explicitly by the model having in mind to compensate rapidly the negative effects that battery voltage drops can have on the closed-loop pressure controller, especially during engine startup phases (e.g. start–stop systems used by microhybrid technology). As regard the alternating component of the rail pressure (or pressure ripple), only a qualitative description has been provided for it in terms of frequency oscillations and maximum amplitude. In what follows we detail the steps of the derivation, identification and experimental validation of the common rail model proposed in [11].

3.1.1 Electrovalve for the Pressure Regulation

The valve used for the regulation of the pressure in the common manifold can be considered as a variable orifice without inertial effects, if the mechanical dynamics of the internal plunger are neglected. Under this assumption, the pressure regulation electrovalve has been modeled as an electrical circuit governed by the equation

$$L\frac{\mathrm{d}i}{\mathrm{d}t} = -Ri + v_d(t), \tag{3.1}$$

where variations of inductance and back-electromotive forces have been furthermore neglected. The parameters L (H) and R (Ω) are the inductance and electrical

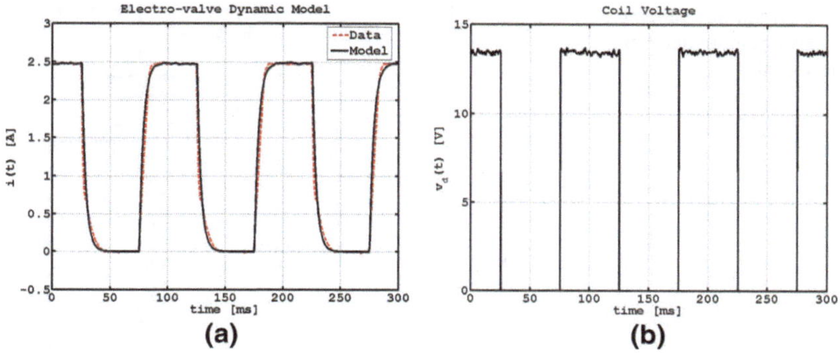

Fig. 3.1 Validation results of the electrovalve dynamic model. Comparison between **a** experimental (*dashed red line*) and model (*red dashed line*) data when **b** the coil voltage is the 10 Hz square wave signal

resistance of the coil, respectively, i (A) is the coil current, and $v_d(t)$ (V) is the instantaneous voltage applied on its poles by the power circuit.

The values of R and L have been identified experimentally (see Table 3.1) through a least-square curve fitting method applied on the error between the model output and experimental data when electrovalve is forced with a battery voltage modulated at 10 Hz. The goodness of the electrovalve model (3.1) under this working condition is confirmed in Fig. 3.1 that shows a very satisfactory agreement between model predictions and experimental data.

3.1.2 Actuation Circuit

Electrovalve is actuated by a single quadrant power circuit as that sketched in Fig. 3.2. Neglecting the voltage drop across the transistor with respect to that battery $V_b(t)$, the potential difference between the coil terminals is

$$v_d(t) = v_c(t)V_b(t), \tag{3.2}$$

where $v_c(t) \in \{0, 1\}$ is a square wave binary signal used to Pulse-Width Modulating (PWM) the control voltage.

The frequency, say f_m, of the PWM signal $v_c(t)$ has been experimentally tuned at 1.5 kHz having been judged a good compromise between a low-output current ripple (that is inversely proportional to f_m) and a limited switching energy losses in power components (that increases with f_m). Furthermore, we point out that this value of frequency modulation guarantees that high frequency harmonics induced by PWM are sufficiently attenuated given that f_m is approximately 30 times the -3 (dB) cutoff

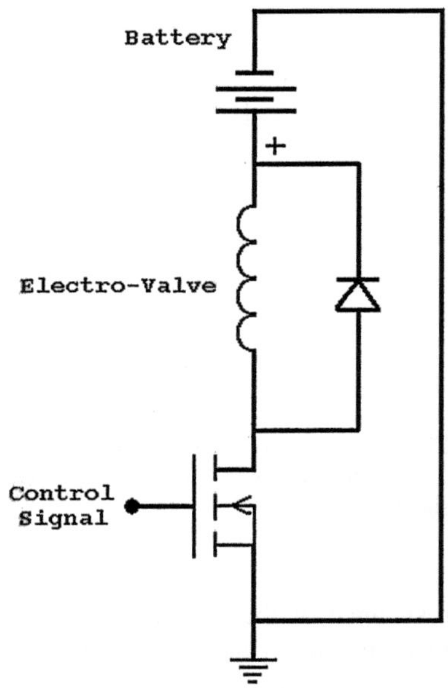

Fig. 3.2 Simplified scheme of the power circuit actuating the electrovalve

Table 3.2 Estimated parameters of the actuation circuit model

Parameter	Unit	Value
a	–	1.0573
b	%	−5.6084

frequency of the electrovalve (i.e. $f_m/f_3 \approx 32$, where $f_3 = 1/(2\pi\tau) = 46\,\text{Hz}$ and $\tau = R/L = 3.46\,\text{ms}$).

Based on these considerations voltage $v_d(t)$ applied on the coil can be well approximated with its average value, say $V_m(t)$, obtained integrating (3.2) over the modulation period $T_m = 1/f_m$, i.e.,

$$V_m(t) = \frac{1}{T_m} \int_{t-T_m}^{t} v_d(\tau)\mathrm{d}\tau = \left(\frac{a\delta + b}{100}\right) V_b(t), \tag{3.3}$$

where $\delta \in [0, 100]$ is the duty cycle percentage of PWM signal $v_c(t)$, and a and b coefficients take into account the nonideal modulation ($a = 1$ and $b = 0$ in the ideal case), whose values, estimated experimentally, are shown in Table 3.2. The battery voltage term $V_b(t)$ in (3.3) is assumed to be constant over a time period T_m.

Experimental values of current ripple obtained when valve is actuated with a duty cycle of $\delta = 30\,(\%)$ and PWM frequency varies $f_m \in \{0.25, 0.75, 1, 1.5, 3\}$ is shown in Fig. 3.3a, whereas the percentage of ripple current with respect to the maximum

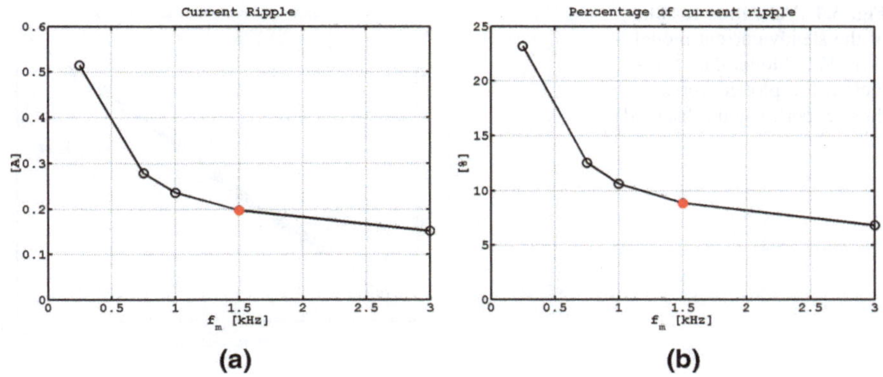

Fig. 3.3 Experimental ripple current observed for a fixed duty cycle of 30 (%) when PWM varies: **a** amplitude; **b** percentage values with respect to maximum coil current. *Red spot* highlights the adopted solution at 1.5 kHz

current flowing in the electrovalve (i.e. $V_b/R \approx 2.2\,\text{A}$, where we assumed $V_b = 12\,\text{V}$), is shown in Fig. 3.3b. A red marker denotes performances at the selected PWM frequency. One can note that model 3.3, evaluated with parameters of Table 3.2, would yield to unfeasible values of actual control voltage applied on the coil (i.e. $V_m < 0$ or $V_m > V_b$) when δ varies in [0, 100]. Therefore, to avoid discrepancies between model and experiments, δ is considered saturated in the interval $[-b/a, \frac{100-b}{a}]$ instead of [0, 100] as in the ideal case.

Replacing (3.3) into Eq. (3.1), the dynamic model of the electrovalve coupled with its power actuator becomes

$$\frac{\mathrm{d}i}{\mathrm{d}t} = -\frac{R}{L}\left[i - f_i(\delta(t), V_b(t))\right], \tag{3.4}$$

where

$$f_i(\delta, V_b) := \frac{V_b}{R}\left(\frac{a\delta + b}{100}\right) \tag{3.5}$$

is the steady response of the current as function of δ (*control variable*) and V_b (*not manipulable input*) as well. A measurement of the battery voltage variable is usually available in the Electronic Central Unit (ECU) and its value used by controllers to quickly adapt main control maps depending on it.

The good agreement between steady model (3.5) and electrovalve with actuation circuit in different steady working conditions is confirmed in Fig. 3.4. Model predicts data with an adjusted coefficient of determination $R^2 = 0.9989$.

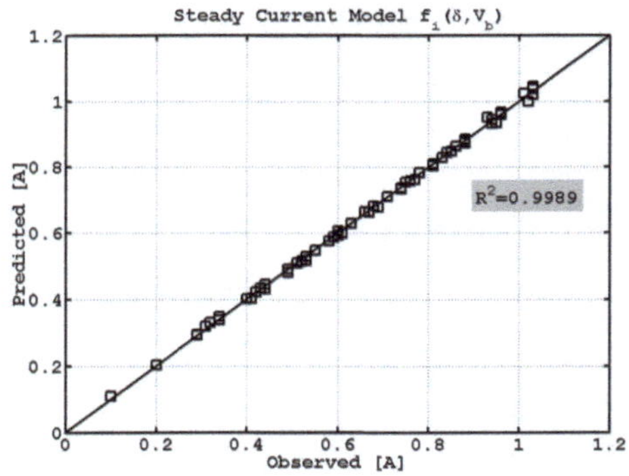

Fig. 3.4 Validation results of the steady current model $f_i(\delta, V_b)$. Model data (predicted) are plotted versus those experimental (observed)

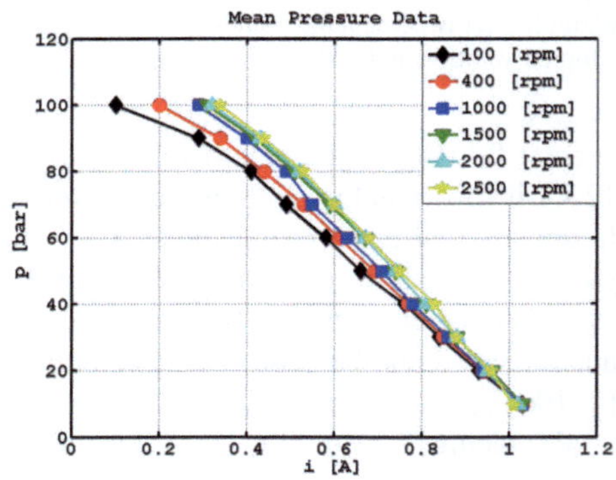

Fig. 3.5 Experimental data of the rail mean pressure $\bar{p}$ versus the electrovalve current i for different HP pump speeds N

3.1.3 Fuel Pressure

A detailed analysis of fluid-dynamics phenomena in such a fuel injection system should be tackled by means of accurate physic mathematical models governing the motion of fluids in a finite pipe (see for example [3]), leading to a set of partial differential equations to be solved numerically by means of complex numerical schemes [12, 13]. Reliable and predictive simulators based on such a detailed models can be obtained and then exploited to analyze in advance the response of the common rail injection system to a given closed-loop pressure control action. Nevertheless, such a detailed modeling is not strictly useful for the design of a model-based controller, whereas a simpler phenomenological model can be well suited for this task.

To this aim, following an experimental approach, we identified a regression model describing the mean pressure, $\bar{p}$, observed in the common rail systems for different values of HP pump speed N ($N := N_e/2$ where N_e is engine speed) and electrovalve current i, when injectors are switched off (see pressure data in Fig. 3.5). According to the data trend analysis the following model has been adopted

Table 3.3 Estimated parameters of the CR pressure model

$c(N)$	Value	$d(N)$	Value
c_0	−97.9354	d_0	113.4366
c_1	−47.5405	d_1	41.8776
c_2	25.0493	d_2	−19.7275
c_3	−4.7962	d_3	3.5148

$$\bar{p} := f_p(i, N) = c(N)i + d(N), \tag{3.6}$$

whose functions $c(N)$ (bar/A) and $d(N)$ (bar) have been first estimated for each N and then fitted[1] with the polynomials

$$c(N) = c_3 \left(\frac{N}{10^3}\right)^3 + c_2 \left(\frac{N}{10^3}\right)^2 + c_1 \left(\frac{N}{10^3}\right) + c_0 \tag{3.7}$$

$$d(N) = \mathrm{d}_3 \left(\frac{N}{10^3}\right)^3 + \mathrm{d}_2 \left(\frac{N}{10^3}\right)^2 + \mathrm{d}_1 \left(\frac{N}{10^3}\right) + \mathrm{d}_0, \tag{3.8}$$

whose coefficients, listed in Table 3.3, have been estimated by a least-square algorithm. Plots of $c(N)$ and $d(N)$ functions are given in Fig. 3.6a and b, respectively.

Mean pressure data yielded by the model (3.6–3.8) are compared with that experimental in Fig. 3.7. Model predicts data with an adjusted coefficient of determination $R^2 = 0.9933$. A three-dimensional (3D) representation of the pressure model (3.6) as function of the electrovalve current and HP pump speed is drawn in Fig. 3.8.

We note that the static pressure is satisfactorily explained by the model up to 80 bar, while for higher values predictive capabilities gets worse. Indeed, pressure slightly deviates from its linear behavior at low current values (see also Fig. 3.5) where the fuel flow is mostly chocked by the electrovalve. It would have been possible to increase model accuracy by introducing a quadratic term in model (3.6), but we prefer to keep the model linear with respect to the state variable i assuming that model mismatches at highest pressure will be compensated via feedback controllers.

Taking into account (3.4–3.6) the complete mean value of CR dynamics are given by

$$\frac{\mathrm{d}i}{\mathrm{d}t} = -\frac{R}{L}i + \frac{V_b}{L}\left(\frac{a\delta(t) + b}{100}\right) \tag{3.9a}$$

$$\bar{p}(t) = c(N)i + d(N). \tag{3.9b}$$

[1] The degree of polynomials has been chosen maximizing the adjusted R^2 statistic index. This index takes into account the number of coefficients used to explain data and it is generally the best measure of fit quality when additional coefficients are added to a model.

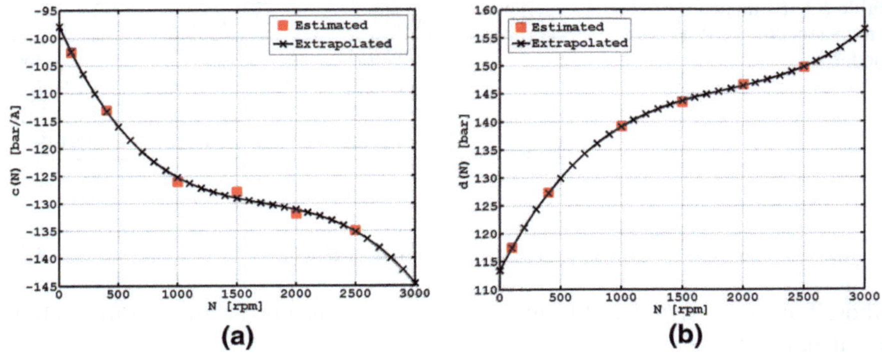

Fig. 3.6 Regression results of **a** $c(N)$ and **b** $d(N)$ parameters in the pressure model

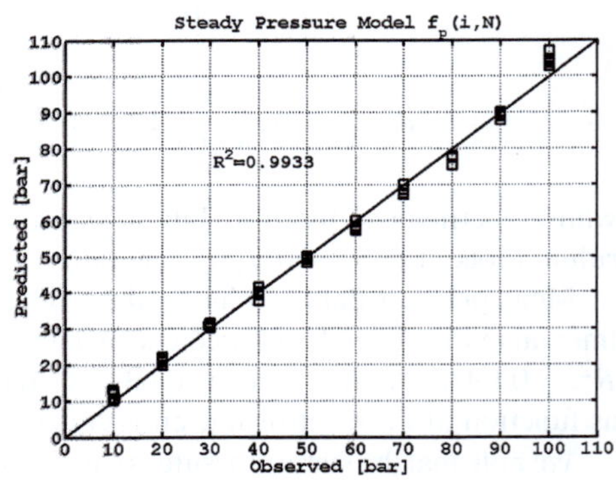

Fig. 3.7 Validation results of the steady pressure model $f_p(i, N)$. Model data (predicted) are plotted versus those experimental (observed)

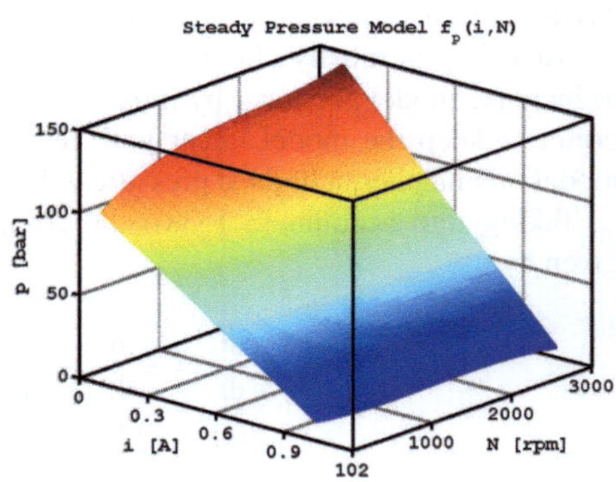

Fig. 3.8 Three-dimensional representation of the steady pressure model in function of the current and HP mechanical speed

3.2 Model Validation Results

The dynamic model (3.4–3.8) developed for the common rail injection system pictured in Fig. 2.12 is experimentally validated in this section. To this aim, three experimental tests at different engine speeds were implemented varying duty cycle with car battery kept under charge at 14.2 V.

The first test (see Fig. 3.9) is related to a sequence of a step responses obtained at the low engine speed, 1000 rpm, when electrovalve is commanded with duty cycle signal depicted in Fig. 3.9b. An overall of the comparison between model and experimental data is shown in Fig. 3.9a, whereas further details are reported in the Fig. 3.9c–h. As shown in these figures, pressure is very well reproduced by the model both during step changes and steady conditions except at 100 bar where model overestimates actual pressure of about 5 %.

In the second test (see Fig. 3.10), the sinusoidal response of common rail pressure augmented with a sequence of three step responses have been investigated at medium engine speed, 2500 rpm. The input signal used to force the system are shown in Fig. 3.10c and d, respectively. As shown in Fig. 3.10a and b a satisfactory agreement between pressure model and experimental data is again confirmed.

Finally, the third test (see 3.11) is referred to a further pressure step response at 5000 rpm obtained forcing the electrovalve with duty cycle the signal shown in Fig. 3.11b. The effectiveness of the model in describing a so wide pressure jump, from 20 to 100 bar, at high engine speed is again confirmed by the comparison between experimental and model data shown in Fig. 3.11a.

3.3 GDI Injectors Model

Since for the design of fundamental automotive control loops, namely Air-to-Fuel Ratio (AFR) control, the control variable to be designed is the mass of fuel to be injected, a relation between this control variable and the duration of injection is fundamental for the integration of the CR-device in the entire engine control system.

In this address, we propose and validate by means of experimental data a simple but effective regression model for the injected mass that takes the following mathematical structure

$$m_{\text{inj}} = T_{\text{inj}} r_1 \sqrt{\Delta P} + (q_1 \Delta P + q_0), \tag{3.10}$$

where m_{inj} (mg) is the actual injected fuel mass, r_1, q_0 and q_1 are the model parameters to be tuned via experiments while $\Delta P_{\text{inj}} = P_{\text{uinj}} - P_{\text{dinj}}$ is the mean of the pressure difference between the upstream injector P_{uinj} (being the fuel pressure into the rail) and the downstream injector P_{dinj} (being the ambient pressure during tests or the in-cylinder pressure when injector is mounted on the engine).

Model structure (3.10) has been derived by analyzing experimental data for different working conditions. More precisely, we have observed that when the average

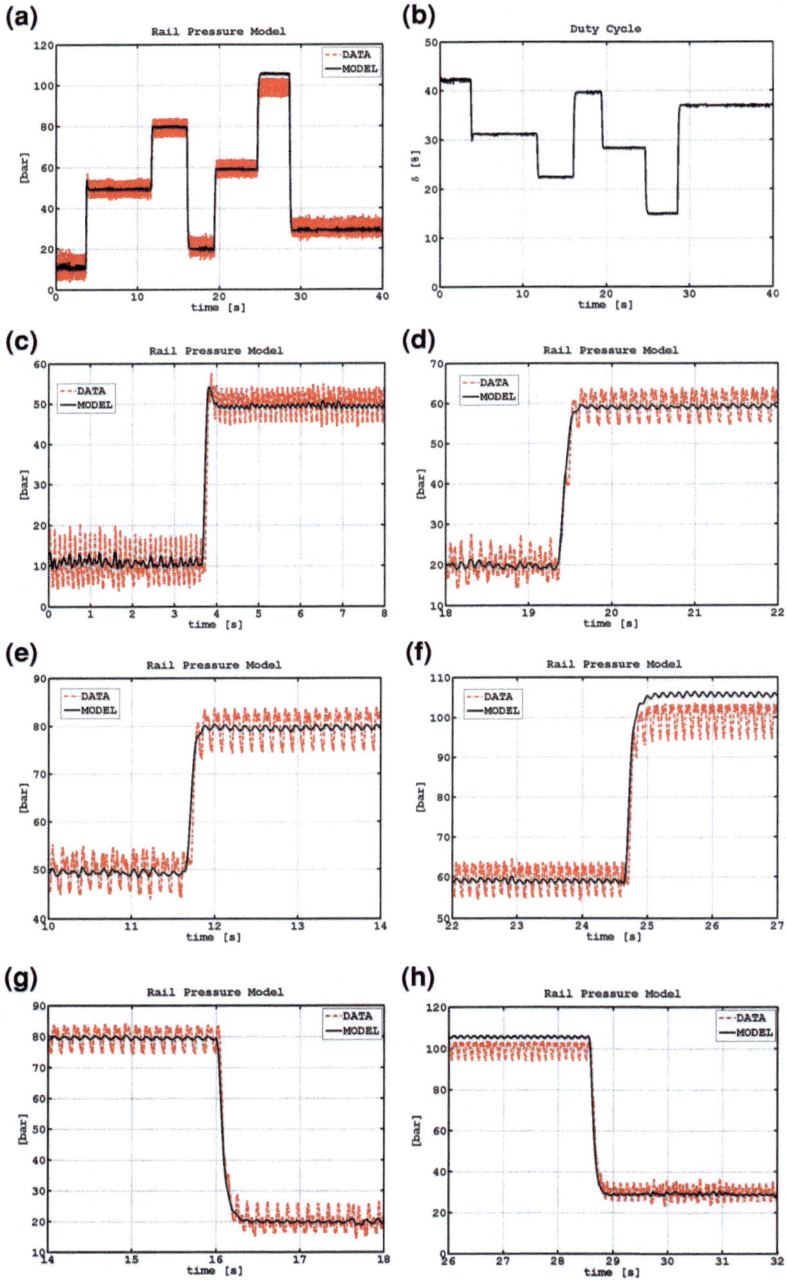

Fig. 3.9 Validation results of common rail pressure model. **a** Overall of the comparison between experimental instantaneous pressure $p(t)$ (*dashed red line*) and mean pressure $\bar{p}(t)$ predicted by the model (*black solid line*) when system is commanded with duty cycle input depicted in **b** and engine speed was 1000 rpm. **c–h** Details of model response at different pressure steps

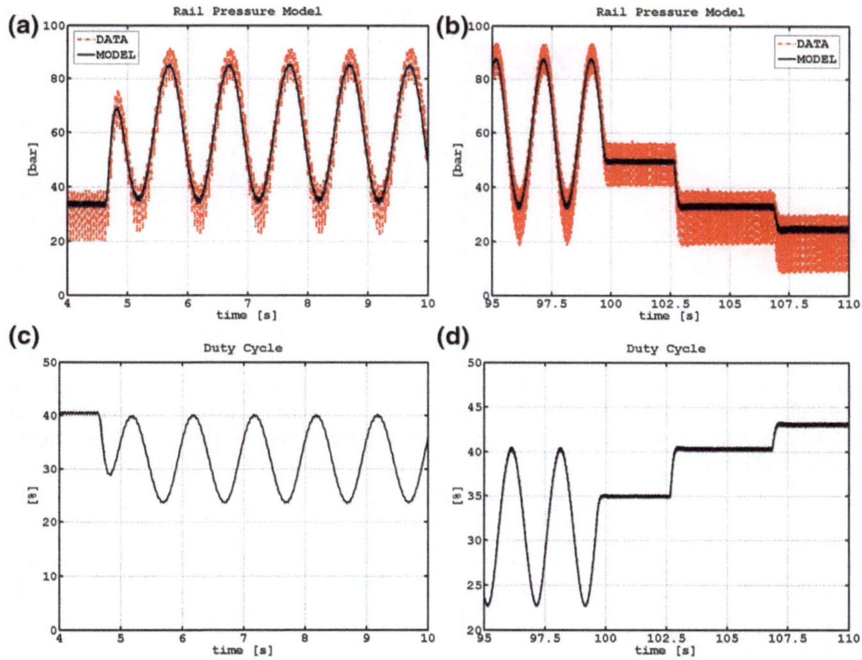

Fig. 3.10 Validation results of common rail pressure model. **a, b** Comparison between experimental instantaneous pressure $p(t)$ (*dashed red line*) and mean pressure $\bar{p}(t)$ predicted by the model (*black solid line*) when system is commanded with duty cycle inputs depicted **c** and **d**, respectively, and engine speed was 2500 rpm

manifold pressure is kept fixed at any constant $\bar{p}$ then the sprayed fuel mass can be well approximated by means of a linear function of the form $m_{\mathrm{inj}} = T_{\mathrm{inj}}r + c$, with r and c being some constants (see Fig. 3.12a). A further experimental investigation has then shown that the coefficients of the straight line, namely r and c, depend on the mean value of common rail pressure as $r(\Delta P) = r_1\sqrt{\Delta P}$ and $c(\Delta P) = q_1 \Delta P + q_0$, respectively. Taking into account the linearity of model (3.10), parameters r_1, q_0 and q_1 can be easily estimated by means of batch least-square algorithm [14]. Numerical values of estimated parameters are listed in Table 3.4.

The effectiveness of injector model (3.10) is confirmed in Fig. 3.12b where model predictions are compared to experimental data for different working points. Model explains data with a very high coefficient of determination equal to $R^2 = 0.99987$.

A mathematical relation for the duration of injection as function of the fuel mass and injection pressure is straightforwardly obtained by inverting the injector model (3.10), that is

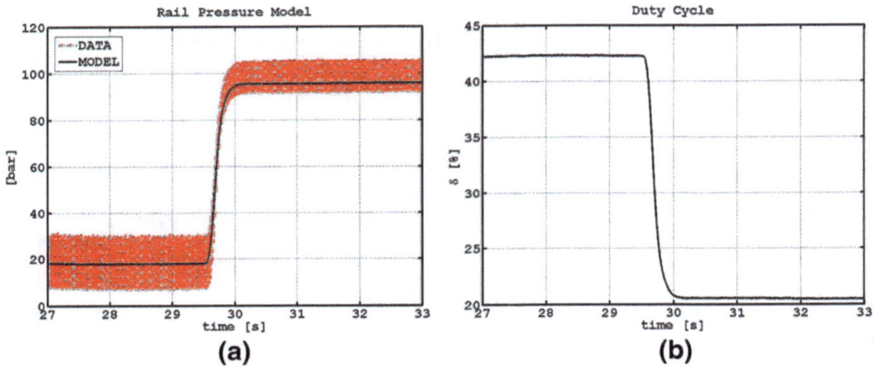

Fig. 3.11 Validation results of common rail pressure model at 5000 rpm. **a** Comparison between experimental instantaneous pressure $p(t)$ (*dashed red line*) and mean pressure $\bar{p}(t)$ predicted by the model (*black solid line*) when system is forced with duty cycle inputs depicted in **b** and engine is speed at 5000 rpm

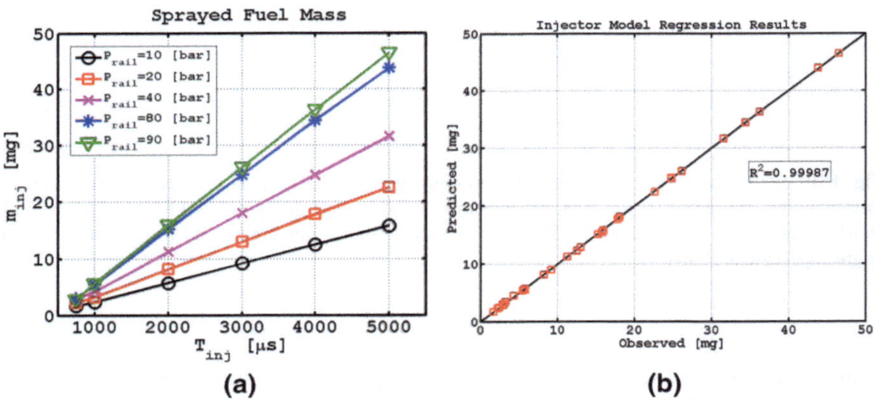

Fig. 3.12 **a** Scatterplot of fuel mass versus duration of injection for different fuel pressure $\bar{p}$. **b** Comparison between measured (observed) fuel mass and that predicted by the model (3.10)

Table 3.4 Estimated parameter for the injector model (3.10)

Parameter	Unit	Value
r_1	mg/μs bar$^{1/2}$	$1.08498 \cdot 10^{-3}$
q_1	mg	$-4.909 \cdot 10^{-2}$
q_0	mg/bar	$-2.99980 \cdot 10^{-1}$

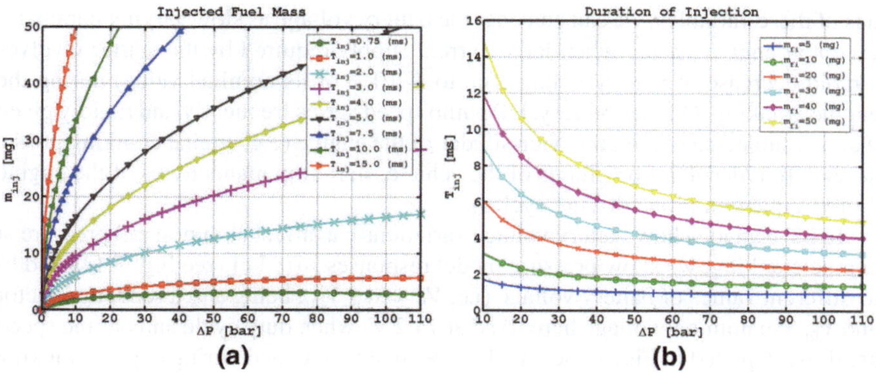

Fig. 3.13 **a** Plots of injected fuel mass model versus injection pressure for different duration of injections. **b** Plots of duration of injection model versus injection pressure for different fuel mass to be injected

$$T_{\text{inj}} = \frac{m_d - q_1 \Delta P - q_0}{r_1 \sqrt{\Delta P}}, \tag{3.11}$$

where m_d is the demanded fuel mass (for example set by the AFR control task) to be sprayed into cylinder at a pressure ΔP. As always happen when device characteristics are derived from experimental data, Eq. (3.11) is valid unless of unavoidable modeling errors.

Plots of injected fuel mass model (3.10) and of duration of injection (3.11) when pressure increases have been shown in the Fig. 3.13a, b, respectively.

3.4 Qualitative Analysis of Plant Disturbances

Since it is important to have a priori information on disturbances acting on the plant to be controlled in closed loop, a qualitative analysis, although roughly, of disturbances deviating common rail pressure from its mean value is tackled in this section. Theoretical analysis supported by experimental observations are presented in what follows.

3.4.1 Battery Voltage and HP Pump Speed

We start to point out that battery voltage, $V_b(t)$, and HP pump speed, $N(t)$, variables appearing in the model Eqs. (3.4) and (3.6), respectively, can be classified as measurable but not manipulable inputs for the common rail plant. (Their measurements are in fact always available in an ECUs.) Moreover, regarding the time variant properties of these variables, we can consider the battery voltage as slow varying parameter except at engine startup when large current supply required by the starter involves a rapid decrease of battery voltage (up to 5–10 % of its nominal value) during the engine cranking. On the contrary, HP pump speed varies frequently and rapidly given that it is subjected to the same vicissitudes of the engine (e.g. engine cranking phase, acceleration/deceleration phases of the vehicle, start-stop manoeuvres of the engine more frequent in urban driving cycles).

To have an idea how battery voltage variations can affect common rail pressure at engine cranking, we show pressure model responses $p(t;\, V_b)$ (see Fig. 3.14a and b) to different values of battery voltage (i.e. $V_b = \alpha_{V_b} V_{\mathrm{bn}}$, being α_{V_b} a reduction factor and V_{bn} the nominal voltage here fixed at 14.2 V) when duty cycle and engine speed are those depicted in Fig. 3.14c and d, respectively, obtained during engine cranking tests of a 2-liter GDI engine. Negative effects are mostly evident during the first 1 s. In particular, when first combustion events occur and electrical starter is turned off (in a neighborhood of 0.5 s, see zoom in Fig. 3.14b), fuel pressure drastically increases with respect to the nominal one (black solid line). As a consequence, considering also that engine is running at low speeds the risk of fuel impingement on cylinder walls increases dangerously.

In order to analyze more strictly the variability of pressure when battery voltage and engine speed changes we calculate the relative-sensitivity functions[2] of the steady pressure function $\bar{p}(\delta, V_b, N)$ with respect to V_b and N, termed $S_{V_b}^{\bar{p}}$ and $S_N^{\bar{p}}$, respectively. By introducing (3.5) in (3.6), the static gain of pressure model writes

$$\bar{p}(\delta, V_b, N) = c(N)\frac{V_b}{R}\left(\frac{a\delta + b}{100}\right) + d(N), \tag{3.14}$$

and sensitivity functions are straightforwardly obtained by definition (3.13), i.e.,

$$S_{V_b}^{\bar{p}}(\delta, V_b, N) = \frac{c(N)}{R}\left(\frac{a\delta + b}{100}\right)\frac{V_b}{\bar{p}(\delta, V_b, N)}, \tag{3.15a}$$

$$S_N^{\bar{p}}(\delta, V_b, N) = \left[c'(N)\frac{V_b}{R}\left(\frac{a\delta + b}{100}\right) + d'(N)\right]\frac{N}{\bar{p}(\delta, V_b, N)}. \tag{3.15b}$$

---

[2] Let $y = f(x; \gamma)$ be a scalar function of $x \in \mathbb{R}^n$ and of $\gamma \in$ parameter. Relative changes in the output $y(x; \gamma)$ due to relative changes of γ can be analyzed by means of relative-sensitivity function that is defined as

$$S_\gamma^f(x; \gamma) = \frac{\Delta y(x)/y(x; \gamma)}{\Delta\gamma/\gamma}, \tag{3.12}$$

where $\Delta y(x) = y(x; \gamma + \Delta\gamma) - y(x; \gamma)$ is the output variation induced by small quantity variation $\Delta\gamma$ of γ parameters. It is easy to get an expression for $S_\gamma^f(x; \gamma)$ by expanding in Taylor's series the $\Delta y(x)$ term, i.e.,

$$S_\gamma^f(x; \gamma) = \frac{\partial f(x; \gamma)}{\partial\gamma}\frac{\gamma}{y(x; \gamma)}. \tag{3.13}$$

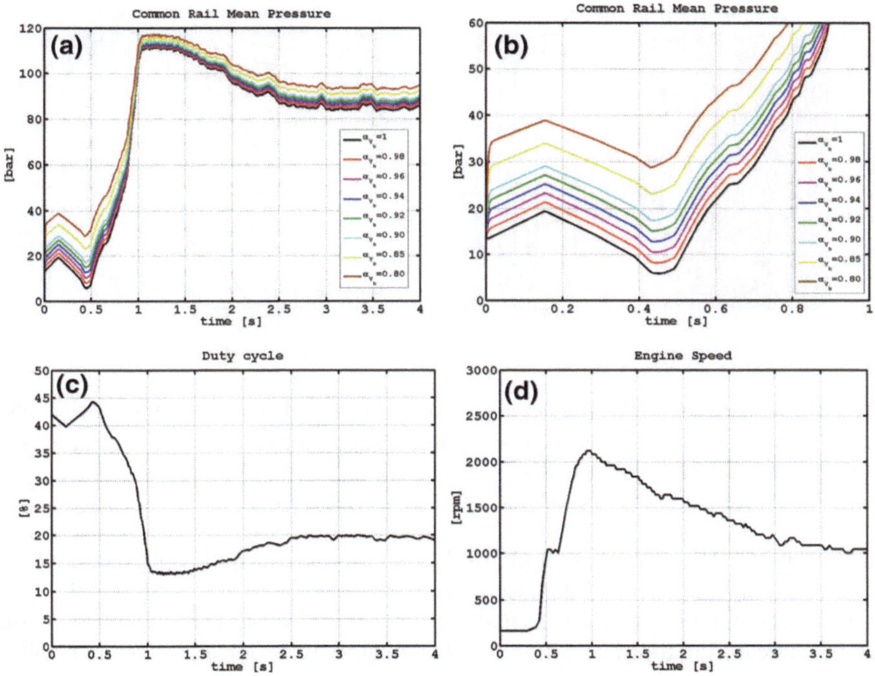

Fig. 3.14 Effects of battery voltage reduction (up to −20 %) on common rail pressure at cranking: **a** simulated pressure for $\alpha_{V_b} \in \{1.00, 0.98, 0.96, 0.94, 0.92, 0.90, 0.85, 0.80\}$; **b** percentage error variations of pressure (zoomed for $t < 1$) with respect to the nominal case $\alpha_{V_b} = 1$; experimental **c** duty cycle and **d** engine speed

Sensitivity functions (3.15a) and (3.15b) are graphically represented by filled contour plots in the Fig. 3.15a and b, respectively. Functions have been evaluated varying δ and N while battery voltage is kept at its nominal value. The values assumed by the function in each area bounded by the black solid isolines read on the colored bar placed on the right side.

As shown in Fig. 3.15a, the blue area on the right side denotes a high sensitivity of the pressure from the battery voltage when duty cycle is approximately close to 45 % independently from engine speed. Just to provide an example, this means that a battery voltage reduction of 5 % can reflect into an increase of common rail pressure of 30 % at low pressure, according to what observed during cranking of the engine when the functioning of common rail injection system just finds in the blue region (see results in Fig. 3.14). The widest region is that orange-red where $S_{V_b}^{\bar{p}}(\delta, V_b, N)$ always assumes values less than 2.

As shown in Fig. 3.15b the sensitivity function with respect to camshaft speed is characterized by a wide green area that extends for engine speed up to 4500 rpm independently from duty cycle. In this region augmented with those smaller cyan

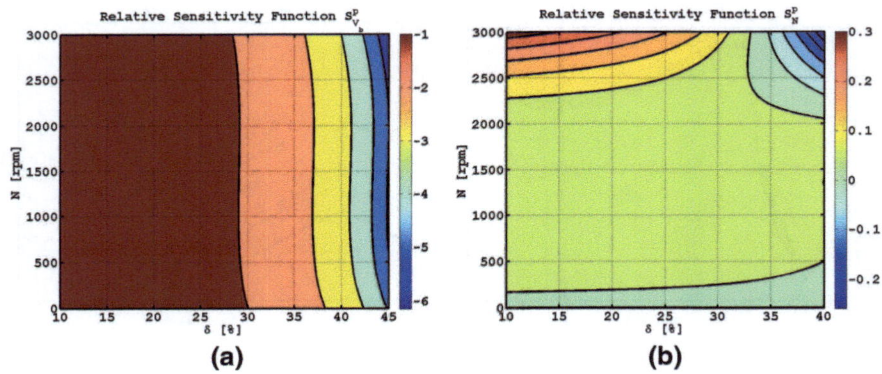

Fig. 3.15 Relative sensitivity functions of pressure to variations in the **a** battery voltage $S_{V_b}^{\bar{p}}$ and **b** HP mechanical speed $S_N^{\bar{p}}$. *Black solid lines* indicate isolines of the function. A constant color is used for each area surrounded by isolines

and yellow, the absolute value of $|S_N^{\bar{p}}(\delta, V_b, N)|$ results always less than 0.1, whilst it never exceeds the value of 0.3 elsewhere.

In order to limit negative effects that such measurable inputs can exert on common rail pressure during transients, it becomes mandatory to design a pressure controller that self adapts its gains in function of the measurable variables V_b and N.

3.4.2 Pressure Ripple

We remind that the model (3.6) describes the average value of the common rail pressure varying the current and HP pump speed when all injectors are disabled. Nevertheless, it is important to have information, even only qualitatively, on the alternating component of pressure, say $\eta(t)$, that inevitably affects the closed -loop chain.

The $\eta(t)$ term takes into account the wave motion of the fuel into the rail induced by the rotational movement of HP pump and of intermittent functioning of the injectors as well. For instance, pressure oscillations originated by the only HP pump are clearly visible in the Figs. 3.9, 3.10 and 3.11.

Let $p(t)$ be the instantaneous value of common rail pressure. From the control viewpoint, the $\eta(t)$ term can be viewed as an additive disturbance to the model output $\bar{p}(t)$, that is

$$p(t) = \bar{p}(t) + \eta(t), \tag{3.16}$$

with $\eta(t)$ decomposed into the sum of two components, i.e.,

$$\eta(t) = \eta_P(t) + \eta_I(t), \tag{3.17}$$

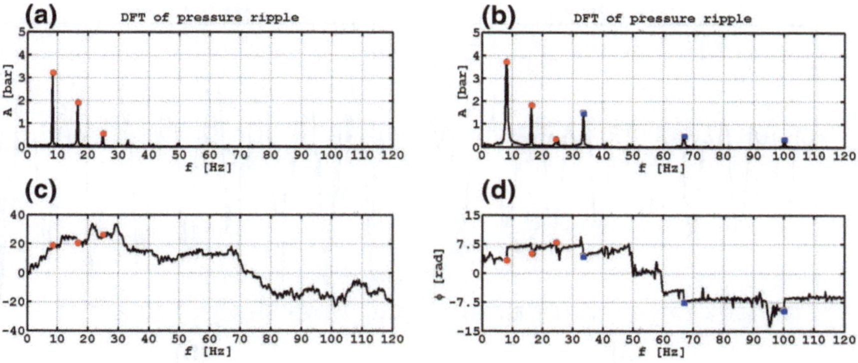

Fig. 3.16 DFTs of pressure ripple observed for a mean pressure $\bar{p} = 60\,\text{bar}$ and an engine speed $N_e = 1000\,\text{rpm}$ when injectors are **a** closed, $T_{inj} = 0\,\text{ms}$ and **b** opened, $T_{inj} = 2.37\,\text{ms}$

where $\eta_P(t)$ and $\eta_I(t)$ are the disturbance originated by the pulsating operations of HP pump and injectors, respectively. The periodic operations of HP pump and injectors imply that the disturbance components in (3.17) are periodic. In fact, for each engine cycle HP pump performs a complete revolution [3] and all injectors are activated sequentially.[4] Hence, fundamental harmonics of the signals $\eta_P(t)$ and $\eta_I(t)$ are respectively located at the frequencies

$$f_P = N/60(\text{Hz}), \tag{3.18a}$$

$$f_I = nN/60(\text{Hz}), \tag{3.18b}$$

with n being the number of injectors supplied by the rail. Note that, the main frequency arising by the injectors is n times that of the HP pump.

Moreover, by the meaning of steady pressure modeled as in (3.6), that is the average of instantaneous pressure $p(t)$ over an engine cycle in absence of injections, i.e., $\eta_I(t) = 0$, and Eq. (3.16) descends that disturbance $\eta_P(t)$ necessarily has a zero mean value. On the contrary, the disturbance $\eta_I(t)$ has a DC component given that when injectors spray a decrease of pressure is observed into the rail. The amplitude of the DC value depends obviously on the fuel mass discharged by the injectors, and so on the duration of the injection command T_{inj} (see Sect. 3.3), for a given injection pressure.

Based on these considerations, disturbances $\eta_{P,I}(t)$ can be approximated by means of the truncated Fourier's series, i.e.,

---

[3] One engine cycle corresponds to a rotation of 720 crankshaft angle degrees that is two engine revolutions.

[4] Each injector is used at least once for engine cycle to spray a given amount of fuel in one or multiple shots.

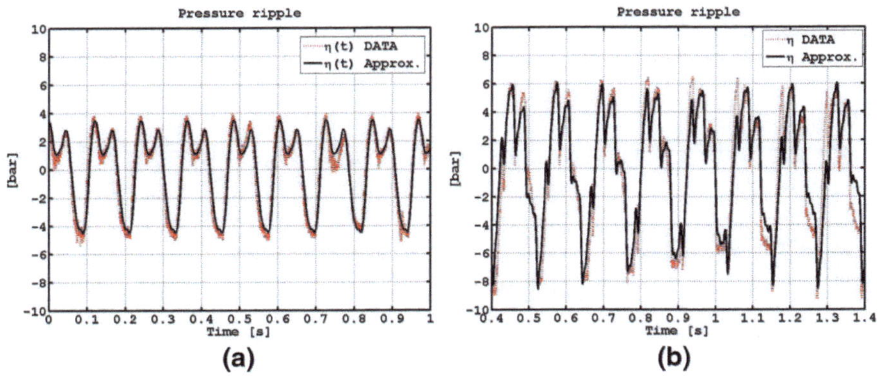

Fig. 3.17 Comparison between experimental (*red dotted line*) and approximated model (*solid black line*) of pressure ripple (3.19) at $\bar{p} = 80\,\text{bar}$, $N_e = 1000\,\text{rpm}$ when injectors are: **a** closed ($T_{\text{inj}} = 0\,\text{ms}$); **b** opened ($T_{\text{inj}} = 2.37\,\text{ms}$)

$$\eta_P(t) \approx +\sum_{k=1}^{3} A_k^P \sin\left(k\frac{\pi N}{30}t + \phi_k^P\right) \tag{3.19a}$$

$$\eta_I(t) \approx -\xi + \sum_{k=1}^{3} A_k^I \sin\left(k\frac{n\pi N}{30}t + \phi_k^I\right), \tag{3.19b}$$

where amplitudes, $A_k^{P,I}$, and phases, $\phi_k^{P,I}$, coefficients, and DC component ξ as well, in general depend on working conditions $\bar{p}$, N and T_{inj}. In particular, one can expect that ξ is mainly influenced by T_{inj} being related to the decrease of mean pressure due to opening of the injectors. Notice that in (3.19) we have assumed that three harmonics are enough to approximate satisfactorily of both disturbances.

In this address, we compute the spectral analysis of pressure ripple for two cases (see Fig. 3.16) related to a mean pressure of 60 bar and a HP pump speed of 500 rpm, when the injectors are switched off and when they are active. In the latter case, the duration of injection has been ($T_{\text{inj}} = 2.37\,\text{ms}$) set for having at 60 bar a stoichiometric air-to-fuel mixture when the air-mass flow rate breathed by the engine is about 30 kg/h for an intake manifold pressure of 750 mbar and an engine speed of 1000 rpm. Moreover, duty cycle has been properly adjusted just to compensate the drop of the mean pressure induced by the opening of four injectors: a decrease of about 4 bar was in fact experimentally observed, ($\xi \approx 4$ bar).

Frequency spectrums of $\eta(t)$ disturbance have been generated via Discrete Fourier Transform (DFT) of residual pressure signals, i.e., $p(t) - \bar{p}$, sampled with a rate of 2 kHz. Amplitude and phase diagrams are shown in the Fig. 3.16a and b. Colored markers highlight the three harmonics considered for the approximated pressure ripple model (3.19). More in detail, the red circles denote the harmonics of $\eta_P(t)$ component, while the blue square indicate those related to $\eta_I(t)$.

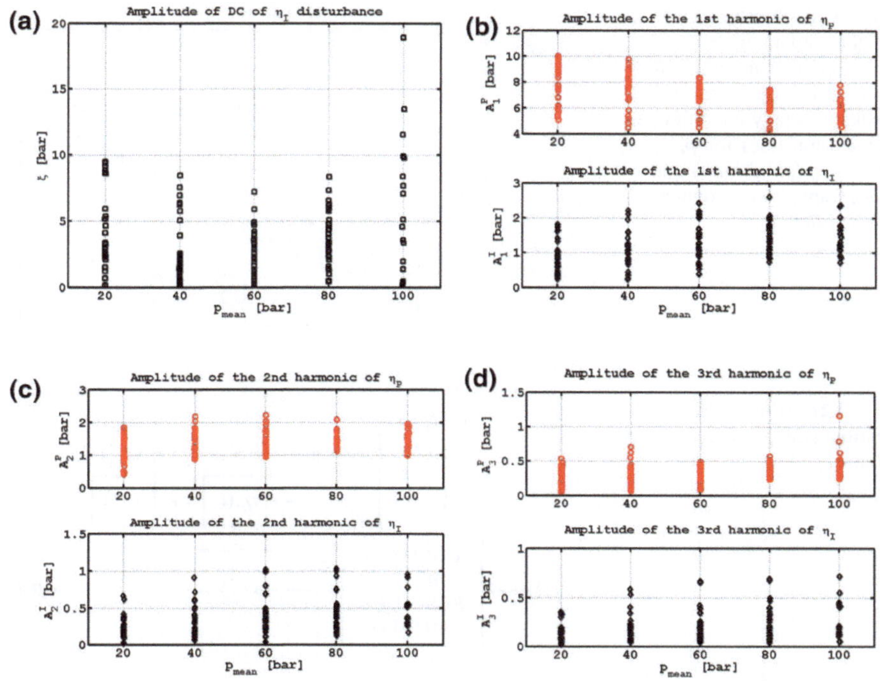

Fig. 3.18 Scatterplots of DC and amplitudes of $\eta_P(t)$ and $\eta_I(t)$ components modeled as in (3.18) versus mean pressure for 90 working points. **a** DC component ξ. **b** Amplitudes A_1^P (*up*) and A_1^I (*bottom*). **c** Amplitudes A_2^P (*up*) and A_2^I (*bottom*). **d** Amplitudes A_3^P (*up*) and A_3^I (*bottom*)

Fig. 3.19 Scatterplot of duration of injections versus mean pressure referred to 90 working conditions

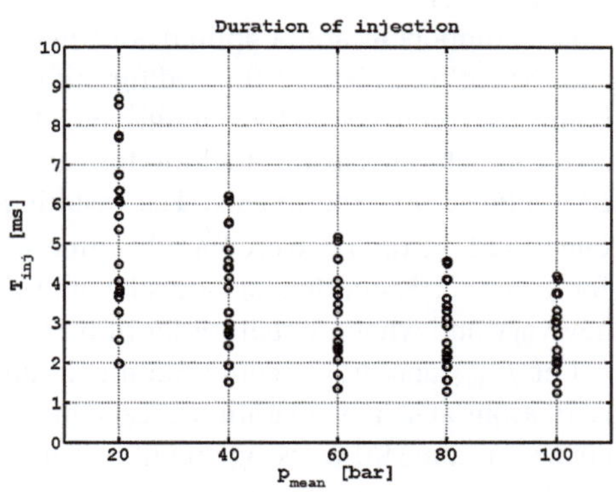

Comparison between experimental pressure ripple and that reconstructed with the functions 3.19 by using actual values of amplitudes $A_k^{P,I}$ and phases $\phi_k^{P,I}$ coefficients as well as the frequencies $kf_{P,I}$, for $k = 1, \ldots, 3$, when injectors are switched off (i.e. the case of $\eta_P(t) \equiv 0$) and when they are active, are shown in the Fig. 3.17a and

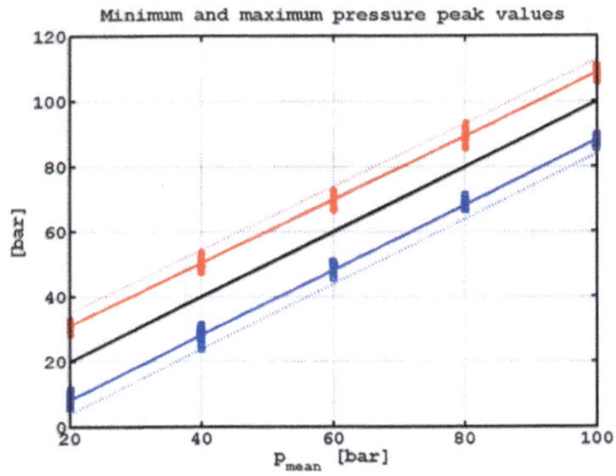

Fig. 3.20 Scatterplots of the maximum $p_{\max}$ (*blue square*) and minimum $p_{\min}$ (*red circle*) pressure peaks at steady conditions. Regression lines for $p_{\max}$ and $p_{\min}$ (*solid lines*) and *upper* and *lower* bounds (*dotted lines*)

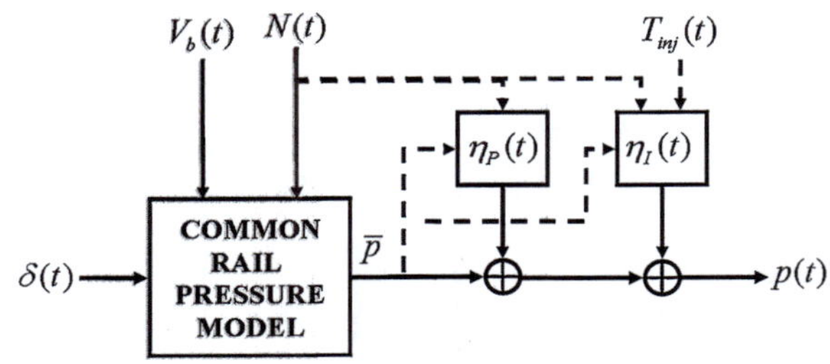

Fig. 3.21 Block schema of common rail plant to put under control

b, respectively. The good agreement between model and experimental data confirms the hypothesis that pressure oscillations can be well reproduced by approximating functions (3.19).

DC component and amplitudes of the first three harmonics of $\eta_P(t)$ and $\eta_I(t)$ disturbances have been estimated for different working conditions, obtained by combining 5 mean pressure by 6 engine speeds by 3 duration of injections (in turn set to have a stoichiometric air–fuel mixture at those speed and mean injection pressure), for a total of 90 experiments. The resulting ninety injection duration are shown in function of the rail pressure for all engine speeds in Fig. 3.19. Scatterplots of ξ, $A_k^{P,I}$ (for $k = 1, 2, 3$) coefficients are shown in Fig. 3.18 in order to have an idea on how they vary and what are their order of magnitudes.

Let $p_{\max}$ and $p_{\min}$ denote the maximum and minimum values, of the pressure oscillations at steady conditions, respectively. Scatterplots of $p_{\max}$ and $p_{\min}$ detected during the same 90 working conditions are shown versus mean pressure in Fig. 3.21 together with the linear regression curves (solid lines), i.e.,

$$p_{\max}(\bar{p}) = 0.97\bar{p} + 11.4 \tag{3.20a}$$

$$p_{\min}(\bar{p}) = \bar{p} - 11.9 \tag{3.20b}$$

and the upper and lower bounds (dotted lines) for $p(t)$ as well, i.e.,

$$p_{\rm ub}(\bar{p}) = p_{\rm max}(\bar{p}) + 4.0 \tag{3.21a}$$

$$p_{\rm lb}(\bar{p}) = p_{\rm min}(\bar{p}) - 4.3. \tag{3.21b}$$

A block scheme for the plant to be controlled in closed loop is shown in Fig. 3.21.

References

1. T. Qiu, Z. Fan, Z. Qi, B. Guo, W. Yin, C. Yu, Study on the influence factors of common rail system's injection mass, in *Proceedings of International Conference on Optoelectronics and Image Processing (ICOIP)*, vol. 2, no. 5, pp. 465–468 (2010)
2. A. de Risi, F. Naccarato, D. Laforgia, Experimental analysis of common rail pressure wave effect on engine emissions. SAE Technical Paper (no. 2005–01-0373) (2005). doi:10.4271/2005-01-0373
3. K. Ahlin, Modelling of pressure waves in the common rail diesel injection system, Ph.D. thesis, University of Linköping, Sweden, 2000
4. M. Corno, F. Casella, S.M. Savaresi, R. Scattolini, Object oriented modeling of a gasoline direct injection system, in *Proceedings of 6th International Modelica Conference* (University of Applied Sciences Bielefeld, Germany, 2008), pp. 83–91
5. M. Corno, S.M. Savaresi, R. Scattolini, E. Comignaghi, M. Sofia, A. Palma, E. Sepe, Modelling, parameter identification and dynamics analysis of a common rail injection system for gasoline engines, in *Proceedings of 17th IFAC World Congress* (Seoul, Korea, 2008). doi:10.3182/20080706-5-KR-1001.01434
6. X.L.J. Seykens, L.M.T. Somers, R.S.G. Baert, Detailed modeling of common rail fuel injection process. MECCA III(2+3), 30–39 (2005), http://www.mate.tue.nl/mate/showabstract.php/6269
7. A.E. Catania, A. Ferrari, M. Manno, Development and application of a complete multijet common-rail injection-system mathematical model for hydrodynamic analysis and diagnostics. J. Eng. Gas Turbines Power **130**(6), 062809 (13 pages) (2008). http://dx.doi.org/10.1115/1.2925679
8. P. Lino, B. Maione, A. Rizzo, A control-oriented model of a common rail injection system for diesel engines, in *Proceedings of the 10th IEEE Conference on Emerging Technologies and Factory Automation (ETFA 2005)* (Catania, Italy, 2005), pp. 563-(7). doi:10.1109/ETFA.2005.1612572
9. C. Gauthier, O. Sename, L. Dugard, G. Meissonnier, Modelling of a diesel engine common rail injection system, in *Proceedings of 16th IFAC World Congress* (Prague, Czech Republic, 2005)
10. A. Balluchi, A. Bicchi, E. Mazzi, A.L. Sangiovanni Vincentelli, G. Serra, Hybrid modelling and control of the common rail injection system. Int. J. Control **80**(11), 1780–1795 (2007)
11. A. di Gaeta, G. Fiengo, A. Palladino, V. Giglio, A control oriented model of a common-rail system for gasoline direct injection engine, in *Proceedings of Joint 48th IEEE Conference on Decision and Control and 28th Chinese Control Conference (CDC/CCC)* (Shanghai, China, 2009). doi:10.1109/CDC.2009.5400211
12. C. Hirsch, in *Numerical Computation of Internal and External Flows–Fundamentals of Numerical Discretization*. Series in Numerical Methods in Engineering, vol. 1, Wiley (1997). ISBN 0-471-92385-0. http://eu.wiley.com/WileyCDA/WileyTitle/productCd-0471923850.html
13. C. Hirsch, in *Numerical Computation of Internal and External Flows–Computational Methods for Inviscid and Viscous Flows*. Series in Numerical Methods in Engineering, vol. 2, Wiley (1998). ISBN 0-471-92452-0. http://eu.wiley.com/WileyCDA/WileyTitle/productCd-0471924520.html
14. L. Liung, *System Identification: Theory for the User*, 2nd edn. (Prentice Hall, Englewood Cliff, 1999). ISBN 0136566952

References

Chapter 4
Synthesis and Experimental Validation of a Fuel Injection Pressure Controller in a Common Rail System

Abstract Electronics has greatly contributed to the development of internal combustion engines. This progress has resulted in reducing environmental degradation, and yet continuing to support improvements in performance. Regarding gasoline engines, a considerable step forward has been achieved by CR technology able to exactly regulate the injection pressure during the entire engine speed range. As a consequence, the injection of a fixed amount of fuel is more precise and it is possible to perform multiple injections for combustion cycle. To obtain this goal a closed loop control must regulate the average of fuel pressure into the rail so to mitigate unavoidable negative effects that the motion of three lobes mechanical pump and multiple injections can have on it. In order to assist the Engine Management System design, through a better performance of GDI engine and the common rail system, in this chapter we present the synthesis and experimental validation of a model-based gain-scheduling controller aimed to regulate the fuel pressure and to track demanded pressure trajectories. By exploiting the simple but effective control oriented model described in the Chap. 3, we get a pressure regulator formed by a closed loop integral action coupled to a feedforward static compensator where both control actions are scheduled in function of the engine speed and battery voltage as well. Rail pressure controller has been experimentally validated for a wide range of working conditions confirming the effectiveness of the proposed control algorithm in regulating the mean value rail pressure independently from engine speed and duration of injection with limited design effort. The resulting controller is simple enough to be effectively implemented in commercial ECUs.

4.1 Introduction

The common rail injection technology has been originally introduced for diesel engines in order to achieve both the reduction of pollutant emissions enforced by international regulations and the improvement of performance required by the

G. Fiengo et al., *Common Rail System for GDI Engines*, SpringerBriefs in Control, Automation and Robotics, DOI: 10.1007/978-1-4471-4468-7_4, © The Author(s) 2013

customers. The key device of this system is the common rail, i.e. a steel manifold where the fuel is kept at high pressure. The electronically controlled high pressure fuel injection system holds an important role concerning both the emission control strategy and the improvement of internal combustion engine performance [4, 9]. High pressure injection allows to finely atomize the fuel spray and to promote fuel and air mixing, resulting in significant combustion improvements [8, 11]. Hence, the control of the injection pressure plays a fundamental role to achieve a better and better engine performance by respecting the current stringent emission regulations.

Recently, this technology has been extended to gasoline engines [3, 10] and significant improvements have been reached adopting more sophisticated and precise control methodologies. As an example, in [2] a controller is designed applying the quantitative feedback theory (QFT) to the closed-loop system. It results in a controller robust to model uncertainties and external disturbances, having moreover a quantitative measure of the robustness achieved. Simulations and experiments on a one-cylinder diesel-dual-fuel engine have shown interesting performance of the proposed controller, compared to the traditional proportional-integral-derivative (PID) controller. In [1] an hybrid model of the Magneti Marelli Powertrain common-rail fuel-injection system for four-cylinders multijet engine has been presented. The hybrid controller is then compared with a classical regulator designed via mean value based approach.

Even though an increasing number of technical papers proposing advanced feedback control loop are available, the control of high pressure in commercial products is still based on open loop strategy based using tables function of pressure. Using the open loop control, injection pressure value is selected directly according to the engine operating conditions. Obviously this approach requires a non negligible time and material resources to identify the pressure mapping, and the resulting maps need to be updated in order to fulfill future emission reduction and fuel economy requirements. On the other hand, when reliable and simple enough models of the pressure rail are available, costs and resources can be reduced by using model-based control strategies.

4.2 Controller Design

The design of a pressure controller which ensures that the mean value rail pressure tracks a reference pressure, say p_d, computed for example by a high level control strategy designed to optimize the engine performance in all operating conditions, according to the following control specifications:

1. zero steady state error ($\bar{p}(\infty) = p_d$);
2. absence of overshoots in the step response ($s_{\%} = 0$, with $s_{\%}$ being the percentage overshoot);
3. step response rise time lower than 250 ms ($t_s < 0.25$ s, with t_s being the rise time);

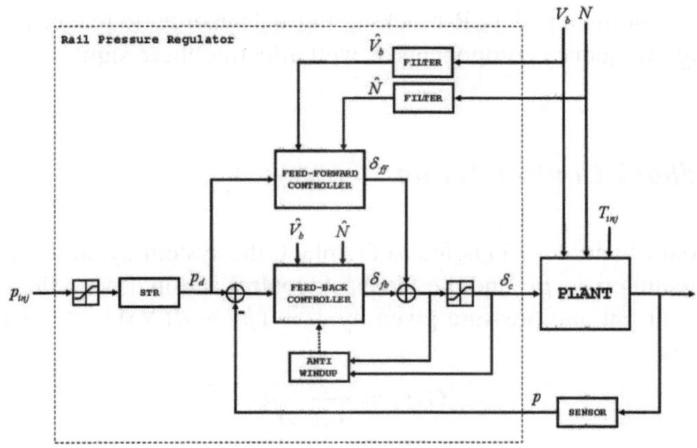

Fig. 4.1 Pressure regulator scheme

4. fast compensation of the not manipulable inputs $N(t)$ and $V_b(t)$;
5. no amplification of the harmonic components of the disturbances η_p and η_I.

Since the plant model strongly depends on the HP pump speed and battery voltage (see Chap. 3), a gain scheduling control algorithm varying the duty cycle (δ) of the pressure regulation electro-valve is here proposed for a fast compensation of their variations (control specification 4). As shown in Fig. 4.1, the control action mainly comprises: (a) a smooth reference trajectory (STR), generating p_d; (b) a feedforward control action δ_{ff}; (c) a feedback control action δ_{fb} based on the difference between the demanded rail pressure and that measured (i.e. $p_e = p_d - p$). In what follows details on the design of each block are given.

The STR-block is mainly a low pass filter to provide a smoother reference pressure to the inner controller. In so doing, the tracking error is limited during fast transients.

4.2.1 Feedforward Control Action

Based on the model (3.9) the feedforward action, δ_{ff}, is designed as the steady state control action needed to keep the mean value pressure at a demanded set point in the nominal case and independently from the working condition. It is then given as:

$$\delta_{\text{ff}}(p_d, \hat{V}_b, \hat{N}) = \frac{100R(p_d - d(\hat{N}))}{a\hat{V}_b c(\hat{N})} - \frac{b}{a}, \tag{4.1}$$

where $\hat{V}_b$ and $\hat{N}$ are the measured battery voltage and pump speed, respectively. Both measurements are obtained by means of proper moving average filters based in the crank-angle domain (FILTER-blocks in Fig. 4.1) in order to reduce measurement noise and high frequency components as well affecting these signals.

4.2.2 Feedback Control Action

When the control action (4.1) is fed to the plant, the system dynamics between the tracking pressure error, p_e, and the feedback control action around the steady state value of the current and pressure given by $\bar{I} = (p_d - d(N))/c(N)$ and $\bar{P} = p_d$, respectively, is

$$G(s) = \frac{K_g}{1 + s\tau}, \tag{4.2}$$

with

$$K_g = \frac{a V_b c(N)}{100R} \tag{4.3a}$$

$$\tau = \frac{L}{R}. \tag{4.3b}$$

The idea is then to design a simple feedback control action δ_{fb} provided by the integral controller

$$C(s) = \frac{K_I}{s}, \tag{4.4}$$

so that the closed loop dynamics match those of a LTI system with transfer function

$$W_d(s) = \frac{1}{(1 + s\tau_d)(1 + s\tau_d')} = \frac{1}{1 + (\tau_d + \tau_d')s + \tau_d \tau_d' s^2}, \tag{4.5}$$

independently from the pump (engine) speed and battery voltage.

We remark that the controller (4.4) has the easiest structure which guarantees zero regulation error in steady state conditions, hence ensuring that condition 1 is fulfilled.

The above control problem can be solved after simple algebraic manipulation by scheduling the control gain K_I as

$$K_I(\hat{V}_b, \hat{N}) = \frac{\tau_d - \tau}{\hat{K}_g(\hat{V}_b, \hat{N})\tau_d^2}, \tag{4.6}$$

with $\hat{K}_g$ being an estimation of (4.3a) and τ_d' chosen as

$$\tau_d' = \frac{\tau \tau_d}{\tau_d - \tau} \qquad \tau_d \in]\tau; \infty[\,. \tag{4.7}$$

Indeed taking into account plant (4.2) and the feedback controller (4.4), the closed loop dynamics are given as:

$$W(s) = \frac{1}{1 + \dfrac{s}{K_g K_I} + \dfrac{\tau s^2}{K_g K_I}}. \tag{4.8}$$

Now to match the dynamics of (4.8) with those of (4.5) the following equalities must be satisfied:

$$\tau_d + \tau'_d = \frac{1}{K_g K_I}, \quad \text{and} \quad \tau_d \tau'_d = \frac{\tau}{K_g K_I} \tag{4.9}$$

From (4.9), it is evident that two degrees of freedom are required to solve the matching problem. Hence, we choose as independent variable K_I and τ'_d. With this choice the matching conditions (4.9) are solved with K_I and τ'_d given in (4.6) and (4.7) respectively.

It is apparent now that only the parameter τ_d has to be selected by practitioners to tune the entire control scheme. More in detail taking into account the control specification 3 we have $\tau_d \lesssim 114$ ms while condition (4.7) imposes $\tau_d > 6.9$ ms. In this work we select this control parameter considering the phase margin of the closed loop system. As shown in Fig. 4.2a the phase margin increases with τ_d while the phase crossover is a decreasing function of τ_d. As a tradeoff between the robustness and tracking performance, we choose $\tau_d = 100$ ms which provides a phase margin $m_\varphi = 88\,°$ at crossover frequency of ω_t=9.6 rad/s. We note with this choice of the τ_d the step response of the reference model does not have overshoots, hence the control specification 2 is ensured.

When the τ_d has been decided it is possible to analyze the closed loop sensitivity function to study if the controller amplifies or not the harmonic components of the disturbance $\eta(t)$. In the case of $\tau_d = 100$ ms, Fig. 4.2b shows that the disturbance attenuation is guaranteed up to about 40 rad/s.

This range of frequency can be extended to higher frequencies choosing a lower value for τ_d, but the risk of inducing instability in the closed-loop system increases since the stability phase decreases with τ_d. For $\tau_d = 2\tau$ the break frequency of sensitivity function could be moved up to 100 rad/s, at risk of an amplification behavior in the range 100–1000 rad/s, with a peak of 1.25 dB at 200 rad/s.

4.3 Injection Time Management

As shown in Sect. 3.4.2 the pressure ripple depends on the motion of the HP pump, disturbance η_P, and the intermittent working of the injectors, disturbance η_I, or in other words on the injection time for a fixed pump speed. In order to generate a realistic disturbance η_I for the testing of the control strategy, the injection time management implemented experimentally has been chosen according to the strategy

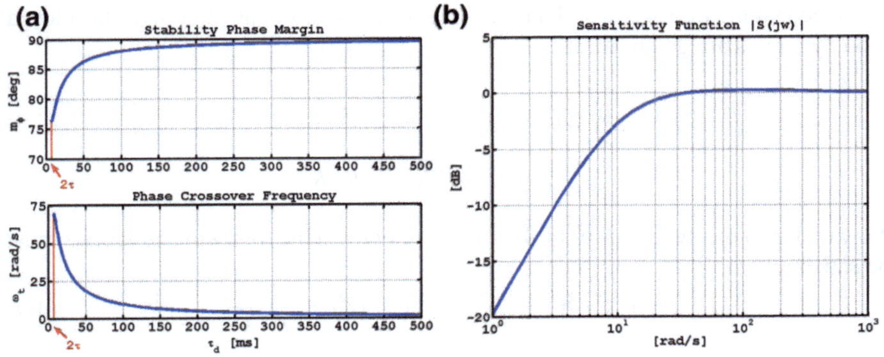

Fig. 4.2 **a** Phase margin in function of τ_d and **b** sensitivity function for $\tau_d = 100\,\text{ms}$

schematized in Fig. 4.3 and described below. Indeed, it is assumed that the fuel mass to be injected in a cylinder, m_d, is that needed to keep the AFR at the stoichiometric value, say S (14.56 for a commercial gasoline). Mathematically it is demanded that $m_{\text{ac}}/(m_d N_{\text{cyl}}) = S$ with m_{ac} being the total of inlet air mass and N_{cyl} the number of cylinders. In order to achieve this control objective often a speed density equation, say f_{sd}, is used to estimate the air mass incoming into cylinder as function of the engine speed, N_{e}, and the intake manifold pressure, P_i. In the rapid prototyping environment, implemented for the testing of the control strategy, both P_i and N_{e} are variables. The pressure variable P_i can be set at different values in order to emulate different engine load while the engine speed variable N_{e} is set as twice the HP pump speed, as it occurs in real vehicles. The HP pump speed is imposed by means of a via a 1.5 kW three-phase induction motor which emulates the motion induced by the internal combustion engine. The speed density equation used is that proposed in [7] for a 2-liters four cylinders GDI engine. Now, since the mean value of CR pressure is on-line computed from the instantaneous pressure, the demanded fuel mass is mapped in an injection time according to the (Eq. 3.11) whose formula is indicated as g in Fig. 4.3.

Remarks

- The simple strategy described above can be assumed as a possible feedforward control strategy to keep the AFR at the stoichiometric value since in the case of GDI engines the phenomenon of wall wetting [6] can be neglected.
- The mean value of rail pressure $\bar{p}(t)$ is computed averaging the instantaneous rail pressure on the engine cycle with a sampling step of 18 crank-angle degrees.

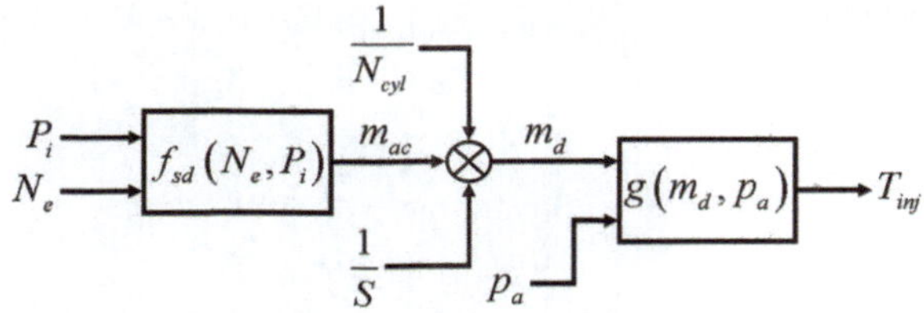

Fig. 4.3 Block scheme for the generation of the time of injection during experimental tests

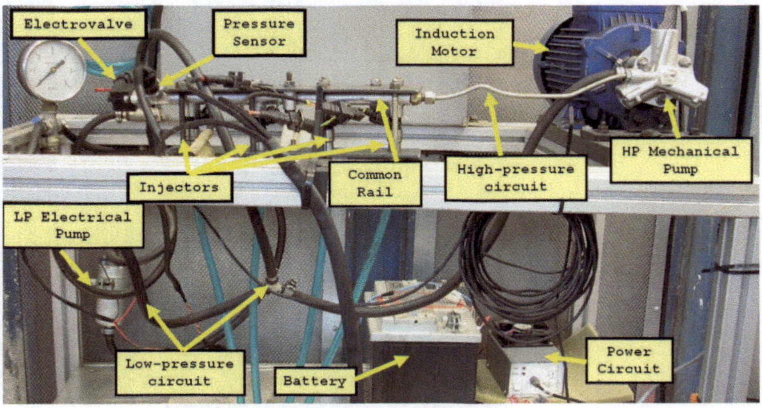

Fig. 4.4 Experimental setup for testing common rail systems

4.4 Experimental Control Setup

Control validation activities have been carried out on the high-pressure injection system for GDI engines shown in Fig. 4.4 whose fundamental components of common rail system are pictured in Fig. 2.12 (see Sect. 2.4).

The pressure sensor (by Bosch) was of a resistive type (Model KV2 BDE) with a time constant of 0.9 ms, a full scale of 140 bar and an accuracy of 1.5 % of measured value. A Hall effect current sensor (Model LTA 50P/SP1 by LEM) with a bandwidth of 100 kHz has been used to measure the coil current. A simple power circuit [5] has been used to actuate control voltage via PWM technique. An oscilloscope (Model TDS-3014 by Tektronix) has been mainly used to record high frequency ripple of the current induced by the PWM at 1.5 kHz.

The HP mechanical pump has been motored by a three phase induction motor of 1.5 kW whose speed was electronically controlled by an AC drive (Model COMMANDER SE 23400220 by Control Techniques) in turn commanded by an external speed reference yielded by the rapid control prototyping (RCP) HW/SW station.

The control scheme presented in Fig. 4.1 has been coded and implemented by means of the RCP station shown in Fig. 4.5, that is based on a multiprocessor system (by dSPACE) equipped with the DS1003 (DSP TMS320C40, 60 Mflops) and

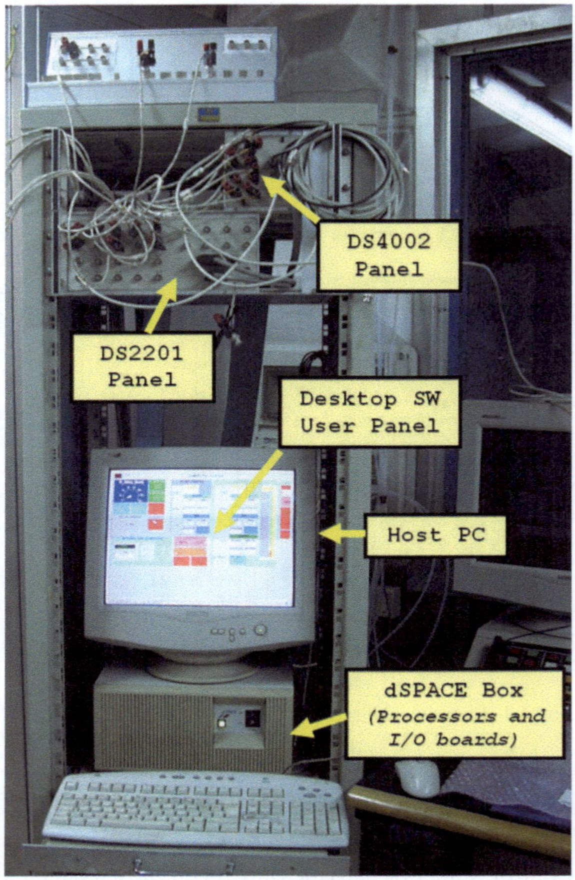

Fig. 4.5 Rapid control prototyping station

DS1004 (DEC Alpha AXP 21164, 600 Mflops) processor boards, and two I/O boards such as the analog DS2201 (20 ch., 12 bit, 30 kHz) and the digital DS4002 (8 ch. CAP/CMP res. 30bit/200 ns, freq. max. 833 kHz) boards. The DSP is programmed in Matlab/Simulink (MathWorks) environment and the experiments are managed and instrumented by a ControlDesk software. The CR pressure control task has been run on the 600 Mflops RISC processor with a sampling time of 1 ms with a turnaround time of about 10 μs.

4.5 Experimental Control Results

In this section the controller designed in Sect. 4.2 is widely tested to show its effectiveness in controlling the mean value of CR pressure, $\bar{p}$, despite the motion of the HP pump and the openings/closings of the injectors which act as disturbances. More

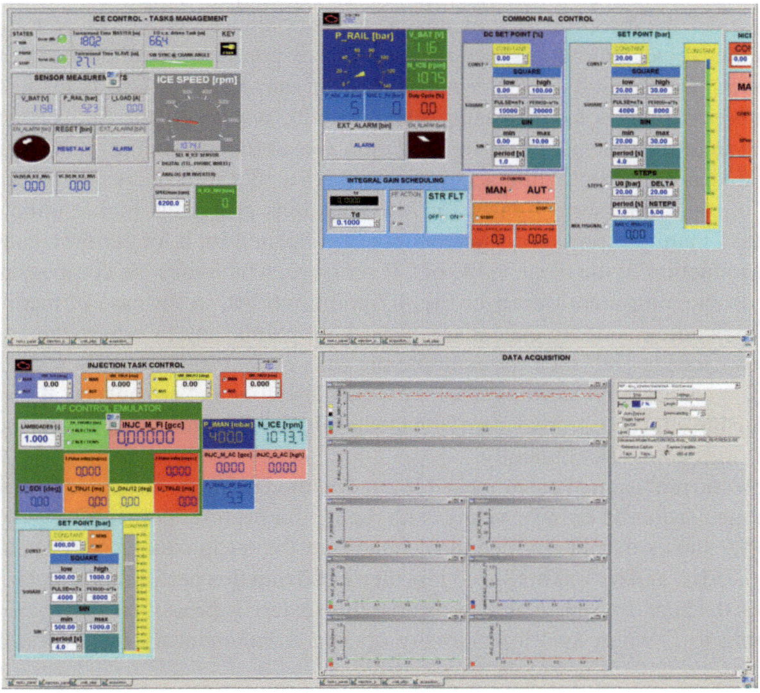

Fig. 4.6 Screen shots of a simple graphical user interface (GUI) designed in ControlDesk to manage control tasks and data capturing

in detail, in order to generate realistic disturbances due to injectors, the injection time management presented in Sect. 4.3 is exploited.

4.5.1 Steady-State Analysis

To illustrate the closed-loop regulation capability in a wide range of operating conditions, in what follows we show that control objectives are achieved for different working points. Precisely, the intake manifold pressure has been varied in the set $P_i \in \mathscr{P}_i \triangleq \{500, 750, 1000\}$ mbar (mimic low, medium and full air load conditions, respectively) while the engine speed takes values in the set $N_e \in \mathscr{N}_e \triangleq \{1000 \cdot k, k = 1, \ldots 6\}$ rpm. For each point in $\Psi \triangleq \mathscr{P}_i \times \mathscr{N}_e$, the reference injection pressure, p_{inj}, has been varied as a sequence of steps taking values in $p_{inj} \in \mathscr{P}_{inj} \triangleq \{20 \cdot k, k = 1 \ldots 5\}$ bar. Closed loop performance have been therefore evaluated in each working point belonging to $\Phi \triangleq \mathscr{P}_i \times \mathscr{N}_e \times \mathscr{P}_{inj}$ both in steady state and during transients from a point to another.

In Fig. 4.7 the injection time expressed in crank-angle degrees ($\theta_{\text{inj}} = 6N_{\text{e}}T_{\text{inj}}{}^{\circ}$) are reported for all the engine speed N_{e} and injection pressure P_{inj} when the mass to be injected m_d and the in-cylinder air mass are those shown in Fig. 4.8a, b, respectively. The injection time is an increasing function with respect to the intake manifold pressure since the fuel mass to be injected increases as function of the intake manifold pressure (see Fig. 4.8a), while according to the Fig. 3.13b it decreases as function of the injection pressure. Hence, for low injections pressures, high engine speed and intake manifold pressure, injectors are kept opened for a longer time causing a greater reduction of the mean value of CR pressure, which must be compensated by the feedback controller. Moreover Fig. 4.7 point out that, in the case of medium ad full load conditions, there exist a set of working points, say Γ, where the angle of injection exceeds 180° (see for example the case $P_i = 1000$ mbar, $N_{\text{e}} = 5000$ rpm and $p_{\text{inj}} = 20$ bar). In those cases there are two injectors that are opened at the same time causing a further decreasing of the mean value of CR pressure if the closed loop control does not work efficiently.

The static precision of the proposed control method has been measured by computing the mean squared error in steady state regime for each set point in Φ (90 points). Figure 4.9 shows that in any operating condition the mean squared error is always below 0.75 bar and for engine speed less then 6000 rpm this performance index never excesses 0.45 bar. In particular we note that this feature is preserved also in the cases belonging to Γ where two injectors are opened simultaneously for a certain time period.

4.5.2 Rise Time Analysis

Considering now the control specification imposed on the rise time of the step response, since it is not easy to measure this quantity experimentally, we introduce the following integral index:

$$I_{\text{rt}} = \sqrt{\frac{1}{T}\int_{t^{\star}}^{t^{\star}+T} (p_{\text{inj}} - \bar{p}(\tau))^2 \mathrm{d}\tau}, \tag{4.10}$$

with $t^{\star}$ being the time instant where a step change occurs, and T being the settling time. Index (4.10) is actually the mean squared error during the transient and evidently depends on the rise time since a greater rise time implies a greater value of the I_{rt}-index. On the other hand, different from the rise time, this performance index can be easily computed numerically by means of an automatic routine considering as T a sufficient large time interval (set to 1.5 s) which guarantees that transient dynamics are vanished.

In order to compare the theoretical performance to those obtained experimentally, the rise time given according with the control design in Sect. 4.2 has been computed

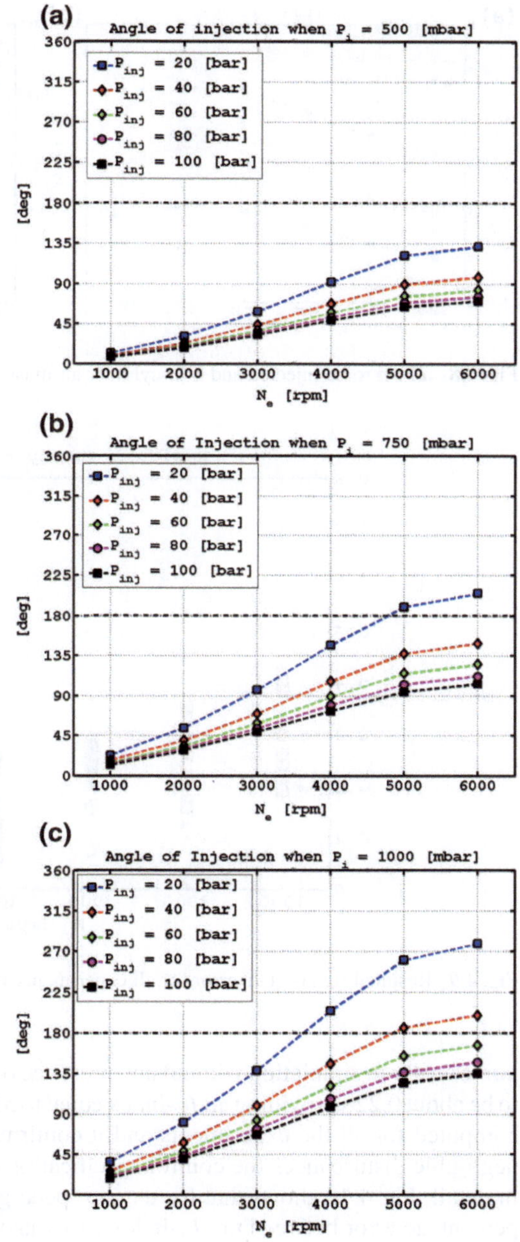

Fig. 4.7 Time of injection, expressed in crank-angle degrees, as function of the engine speed for different injection pressure when the intake manifold pressure P_i is: **a** 500 mbar; **b** 750 mbar and **c** 1000 mbar

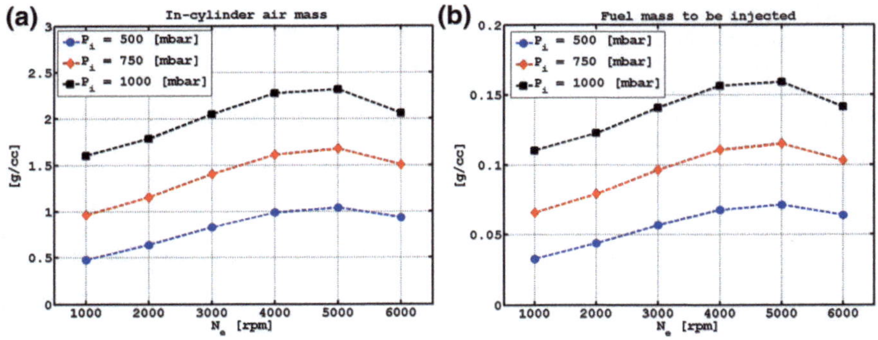

Fig. 4.8 **a** Fuel to be injected and **b** in-cylinder air mass for each working point in $\mathscr{P}_i \times \mathscr{N}_e$

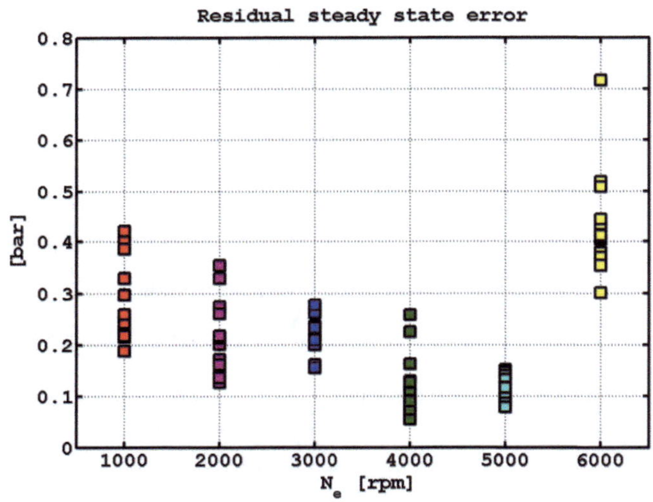

Fig. 4.9 Residual steady state error for all experiments as function of the engine speed

numerically in simulation without any presence of the disturbance η and it was found to be about 0.22 s providing an I_{rt}-index equal to 5 bar. Figure 4.10 shows the I_{rt}-index computed for all the experiments and it confirms that even in the presence of non negligible disturbances the control specification on the rise time is achieved. More in detail, Fig. 4.10 shows that for engine speed grater than 2000 rpm the maximum percentage error between the I_{rt}-index target and those computed experimentally is at most 10 % and for engine speed belonging to 4000, 5000, 6000 rpm there are some cases where the achieved I_{rt}-index is smaller than that required. Only for low engine speed performance deteriorate slightly with a maximum percentage errors of about 15 % at 2000 rpm and 30 % at 1000 rpm.

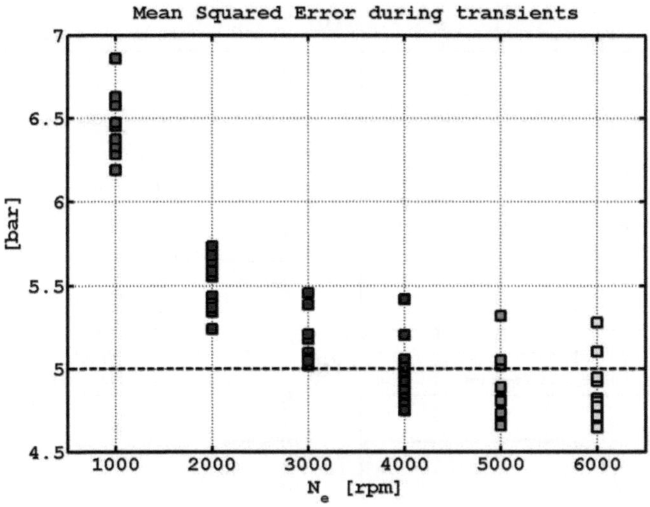

Fig. 4.10 I_{rt}-index for all experiments as function of the engine speed

To confirm furthermore the effectiveness of the proposed control strategy, in what follows some time histories, one for each engine speed, are reported and briefly analyzed for the sake of brevity. As shown in Fig. 4.11a, in the case of $N_e = 1000$ rpm the closed loop dynamics are slightly slower then those demanded, hence the I_{rt}-index increases as discussed before. Nevertheless the steady state error is below 0.45 bar as shown in Fig. 4.9 and zero overshoot is achieved. Tracking performance increase enormously for greater engine speed as shown for example in Fig. 4.11b and c for engine speed of 2000 and 3000 rpm, respectively. In this cases the steady state errors are below 0.4 bar and they are still achieved without overshoots.

For high engine speed more complex cases are in what follows analyzed. Figure 4.11d shows the regulating performance in the case of an engine speed of 4000 rpm, intake manifold 1000 mbar and the injection pressure equal to 20 bar, while Fig. 4.11e reports the case when the engine speed is 5000 rpm, intake manifold 1000 mbar and the injection pressure equal to 40 bar. These working conditions are particulary difficult to be tackled since, as shown in Fig. 4.7b, c, the angle of injection excess 180°, hence two injectors work simultaneously inducing at the same time a greater reduction of the mean value pressure and more complex oscillating dynamics of the pressure waves in the rail. Despite of the presence of such disturbances the mean value pressure dynamics still match well those of the STR output both in steady state and during transient. Finally, the case of high engine speed ($N_e = 6000$ rpm and high injection pressure ($p_{inj} = 100$ bar) is reported in Fig. 4.11f. Also in this case the mean rail pressure tracks the demanded command with steady state error less than 0.75 bar, a percentage tracking error less than 10 % without overshoots.

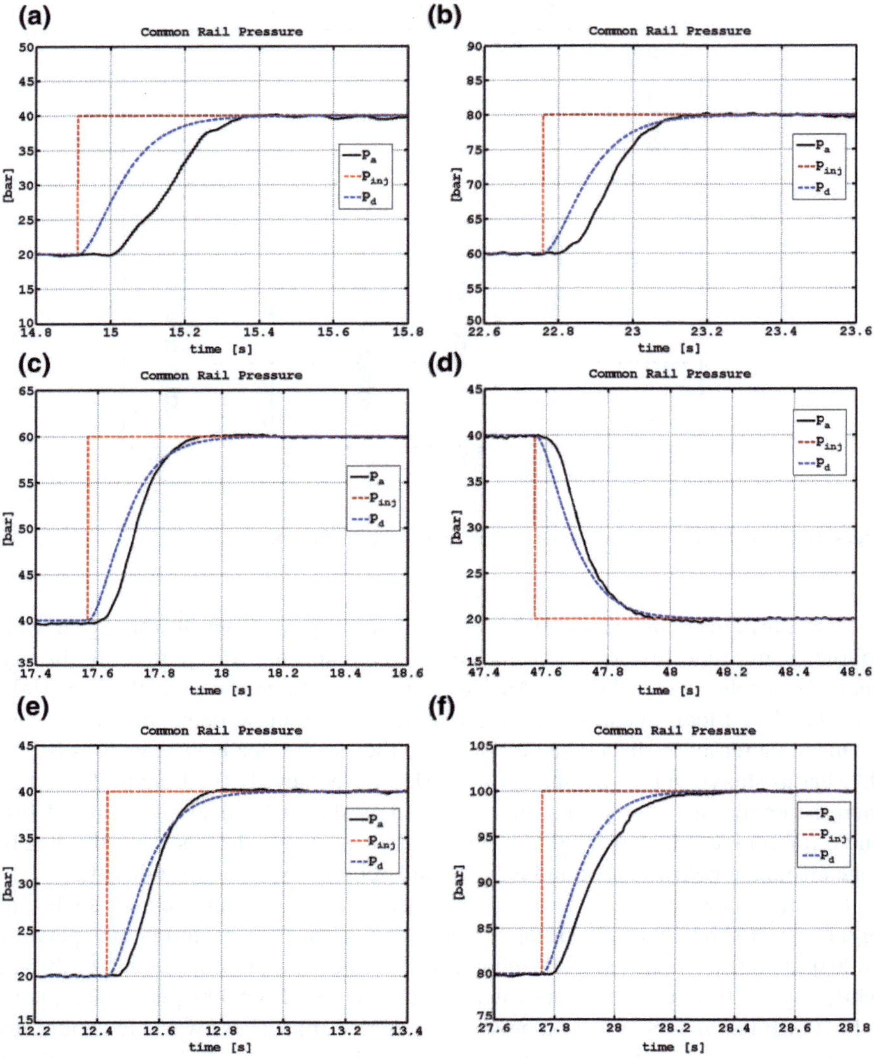

Fig. 4.11 Experimental control results. Mean value CR pressure (*black solid line*) $\bar{p}$, demanded p_{inj} (*red dashed line*), demanded filtered pressure p_d (*blue dashed line*) when: **a** $N_e = 1000$ rpm, $P_i = 750$ mbar and $p_{\text{inj}} = 40$ bar; **b** $N_e = 2000$ rpm, $P_i = 500$ mbar and $p_{\text{inj}} = 80$ bar; **c** $N_e = 3000$ rpm, $P_i = 500$ mbar and $p_{\text{inj}} = 60$ bar; **d** $N_e = 4000$ rpm, $P_i = 1000$ mbar and $p_{\text{inj}} = 20$ bar; **e** $N_e = 5000$ rpm, $P_i = 1000$ mbar and $p_{\text{inj}} = 40$ bar; **f** $N_e = 6000$ rpm, $P_i = 750$ mbar and $p_{\text{inj}} = 100$ bar

Remarks

- We remark that zero overshoot has been attained in closed-loop not only for the cases considered in Fig. 4.11 but for all the operating conditions. Hence, also the control specification on absence of overshoot is experimentally confirmed.
- The duty cycle for each case analyzed above is reported in Fig. 4.12 confirming the feasibility of control action. In particular, absence of saturation of the control action have been achieved in any working condition both during transients and in steady state, but further time histories are not reported for the sake of brevity.
- The actual pressure for each working point previously considered is shown in Fig. 4.13. The presence of the bounded ripple is due to disturbance η that persistently acts on the plant. As it is apparent the frequency of the oscillation increases as function of the engine speed as predicted by the model in Chap. 3. Nevertheless, the reduction of such oscillations is out of the scope of the present book.

  Regarding to the residual pressure, Fig 4.14 shows the peak to peak amplitude of the oscillation around the mean value rail pressure. For each working point in Φ, the residual pressure, say r_p, has been computed considering in steady state enough peaks of pressure above the mean value, say P_{u_k}, and peaks under the mean value, say P_{l_k}, $k = 1 \ldots M$ with M being the number of peaks considered for the residual pressure computation. Hence, the amplitude of the pressure oscillation has been computed as

$$r_p = \frac{1}{M}\left(\sum_{k=1}^{M} p_{u_k} - \sum_{k=1}^{M} p_{l_k}\right). \tag{4.11}$$

  As shown in Fig. 4.14 the residual pressure is limited in the range of 6–22 bar and has an increasing mean trend as function of the engine speed for each intake manifold pressure. Finally, we note that for each working point the same order of magnitude of the residual pressure has been obtained in open-loop but results are not reported here for the sake of brevity. Hence, the control specification on the non amplification of the harmonic components of the disturbance η is satisfied.

4.5.3 Tracking Performance

Finally, we consider the case of the tracking of a time varying reference signal. Indeed, we test the control performance for different conditions when the demanded injection pressure is a sinusoidal wave with period 2 s, amplitude 30 bar and bias 60 bar and the engine speed is kept at 2500 rpm. We note that with this choice the reference input changes from 30 to 90 bar and viceversa in about 20 engine cycles that can be considered. Some details on the experimental results are provided in what follows.

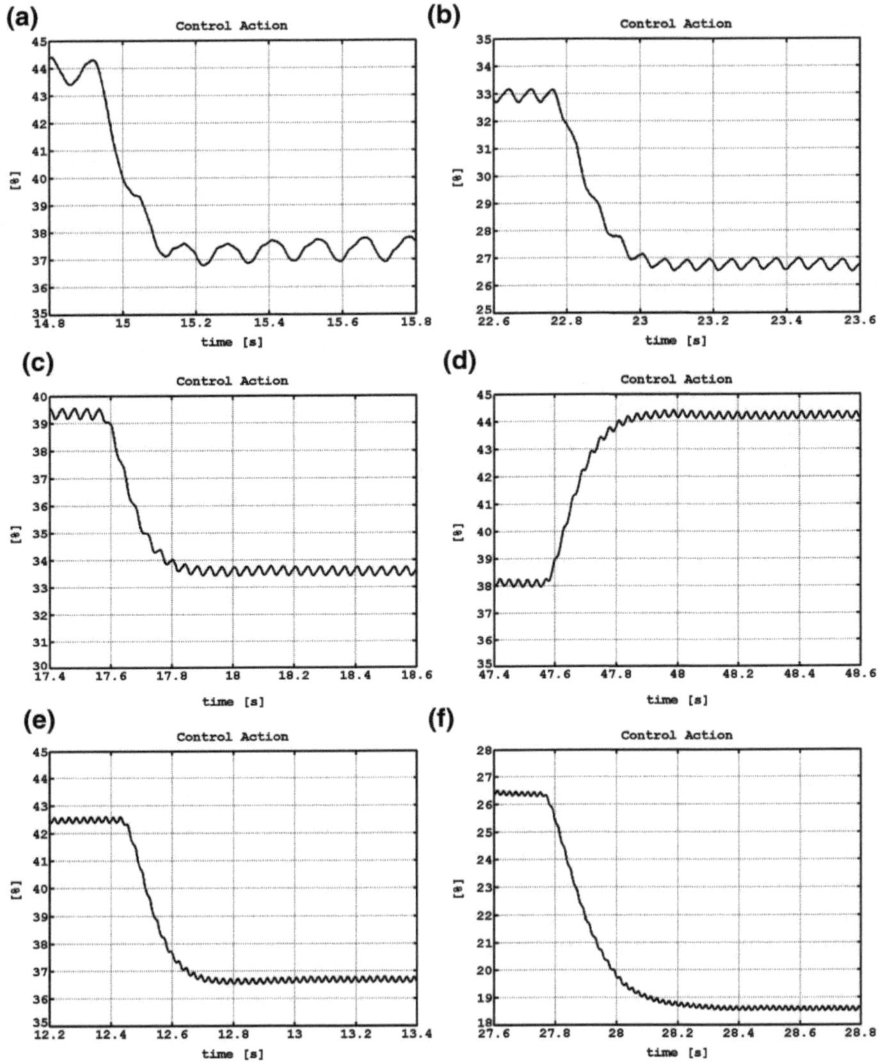

Fig. 4.12 Experimental control action when: **a** $N_e = 1000\,\text{rpm}$, $P_i = 750\,\text{mbar}$ and $p_{inj} = 40\,\text{bar}$; **b** $N_e = 2000\,\text{rpm}$, $P_i = 500\,\text{mbar}$ and $p_{inj} = 80\,\text{bar}$; **c** $N_e = 3000\,\text{rpm}$, $P_i = 500\,\text{mbar}$ and $p_{inj} = 60\,\text{bar}$; **d** $N_e = 4000\,\text{rpm}$, $P_i = 1000\,\text{mbar}$ and $p_{inj} = 20\,\text{bar}$; **e** $N_e = 5000\,\text{rpm}$, $P_i = 1000\,\text{mbar}$ and $p_{inj} = 40\,\text{bar}$; **f** $N_e = 6000\,\text{rpm}$, $P_i = 750\,\text{mbar}$ and $p_{inj} = 100\,\text{bar}$

Figure 4.15 shows the tracking performance in the case of absence of injections (the disturbance $\eta_I \equiv 0$). The delay and the attenuation of the mean value CR pressure is mainly due to the presence of the STR filter whose output is the actual signal to be tracked in closed loop.

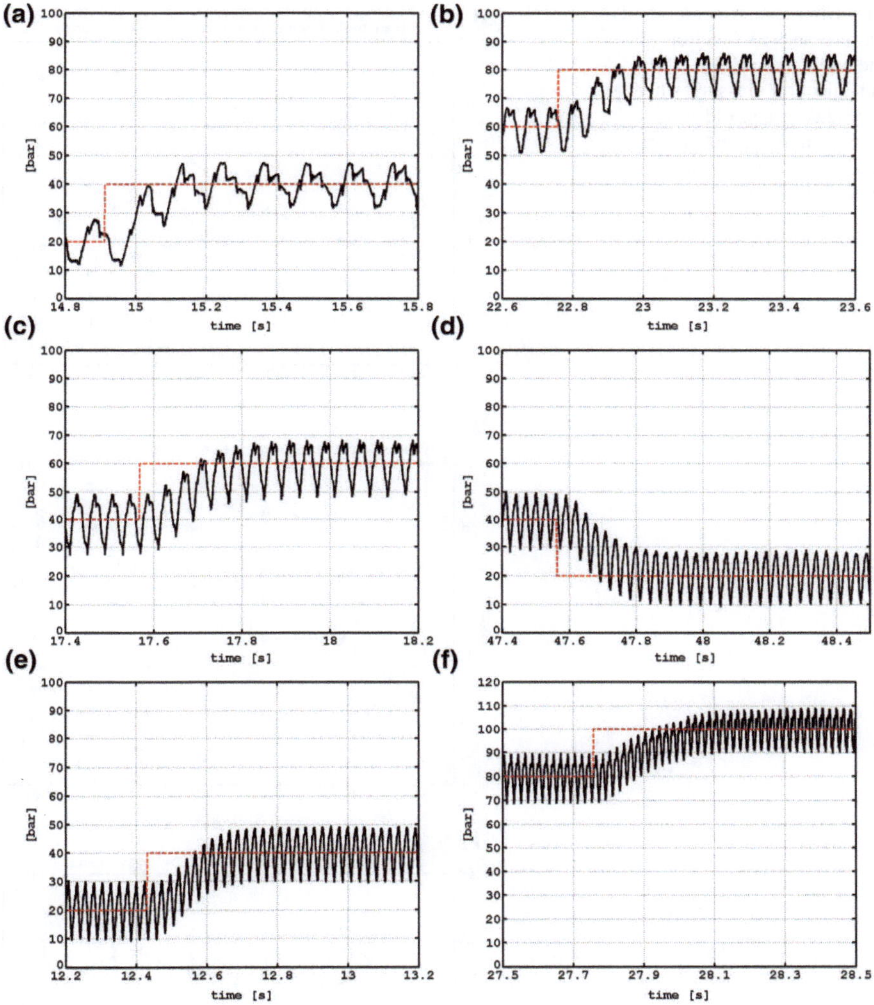

Fig. 4.13 Experimental CR pressure (*black solid line*) and demanded pressure (*red dashed line*) when: **a** $N_e = 1000$ rpm, $P_i = 750$ mbar and $p_{inj} = 40$ bar; **b** $N_e = 2000$ rpm, $P_i = 500$ mbar and $p_{inj} = 80$ bar; **c** $N_e = 3000$ rpm, $P_i = 500$ mbar and $p_{inj} = 60$ bar; **d** $N_e = 4000$ rpm, $P_i = 1000$ mbar and $p_{inj} = 20$ bar; **e** $N_e = 5000$ rpm, $P_i = 1000$ mbar and $p_{inj} = 40$ bar; **f** $N_e = 6000$ rpm, $P_i = 750$ mbar and $p_{inj} = 100$ bar

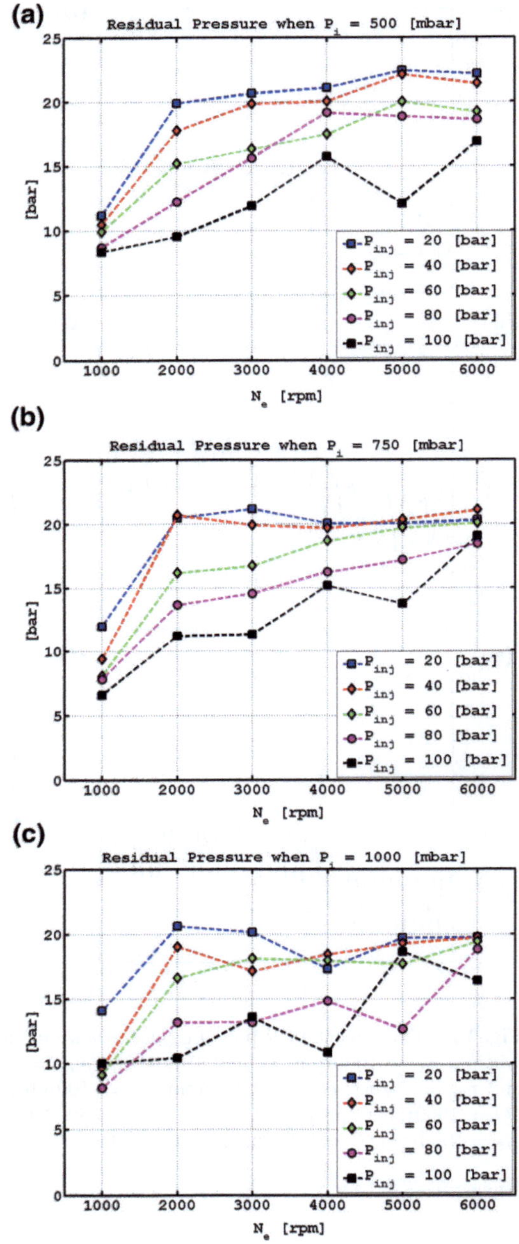

Fig. 4.14 Experimental residual pressure as function of the engine speed for different injection pressure when the intake manifold pressure is: **a** $P_i = 500\,\text{mbar}$; **b** $P_i = 750\,\text{mbar}$ and **c** $P_i = 1000\,\text{mbar}$

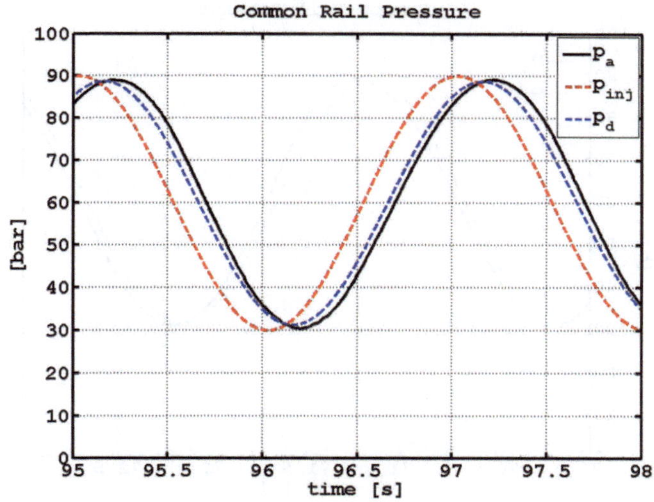

Fig. 4.15 Experimental control results. Mean value CR pressure $\bar{p}(t)$ (*black solid line*), demanded pressure $p_{\text{inj}}(t)$ (*red dashed line*) and filtered demanded pressure $p_d(t)$ (*blue dashed line*) when engine speed is $N_e = 2500\,\text{rpm}$ and demanded injection pressure is a sinusoidal wave with period 2 s, amplitude 30 bar and bias 60 bar when *no fuel injections* occur

For emulating the disturbances introduced by a vehicle driver during rapid tip-in/tip-out manoeuvres, we have tested the control action when the intake manifold pressure varies as a sinusoid with period 2 s and oscillating between 400 mbar and 1000 mbar. In Fig. 4.16 is shown the experimental results when the injection fuel mass, computed according with the procedure detailed in Sect. 4.3, is completely sprayed with a single injection per engine cycle, while the tracking performance in the case of double injection per engine cycle is reported in Fig. 4.17. More in detail, when this injection strategy is used an equal amount of fuel mass is sprayed during the first and the second injection with a separation crank-angle between the two injections of 30°. In both cases the control performances are very satisfactory and in particular the presence of a double injection does not affect the closed loop mean value dynamics.

For all the cases described above, Fig. 4.18 shows the time histories of the actual CR pressure as well as the output of the STR filter. As apparent the mean value pressure is around the demanded injection pressure independently from absence or the presence of injections and different number of the injections per engine cycle. Hence, the control objective is still guaranteed also for these more complex manoeuvres. We note that also for these cases the control input is in the admissible range, but results are not reported here for the sake of brevity.

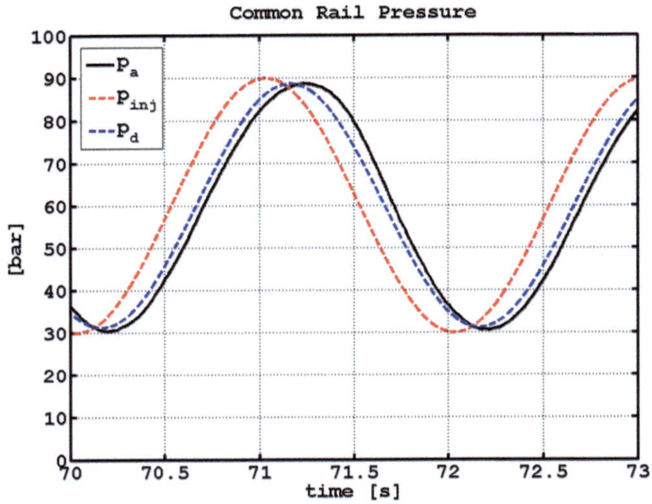

Fig. 4.16 Experimental control results. Mean value CR pressure $\bar{p}(t)$ (*black solid line*), demanded pressure $p_{\text{inj}}(t)$ (*red dashed line*) and filtered demanded pressure $p_d(t)$ (*blue dashed line*) when engine speed is $N_e = 2500\,\text{rpm}$ and demanded injection pressure is a sinusoidal wave with period 2 s, amplitude 30 bar and bias 60 bar in the case of *single injection* per engine cycle when the intake manifold pressure $P_i(t)$ changes sinusoidally with period 2 s, amplitude 300 mbar and bias 700 mbar

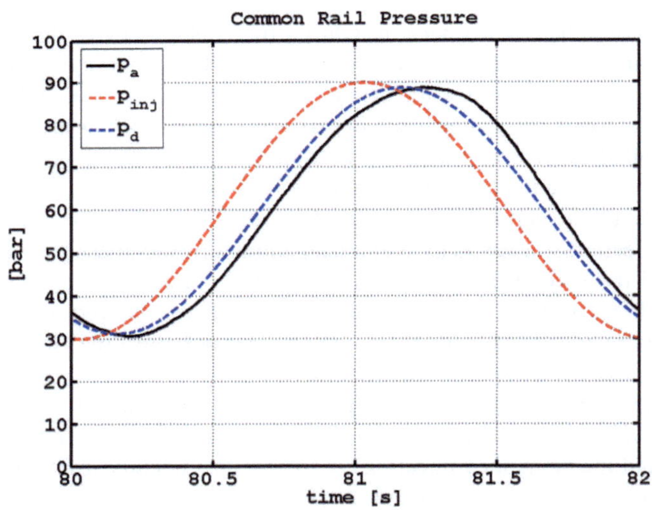

Fig. 4.17 Experimental control results. Mean value CR pressure $\bar{p}(t)$ (*black solid line*), demanded pressure $p_{\text{inj}}(t)$ (*red dashed line*) and filtered demanded pressure $p_d(t)$ (*blue dashed line*) when engine speed is $N_e = 2500\,\text{rpm}$ and demanded injection pressure is a sinusoidal wave with period 2 s, amplitude 30 bar and bias 60 bar in the case of *double injection* per engine cycle when the intake manifold pressure $P_i(t)$ changes sinusoidally with period 2 s, amplitude 300 mbar and bias 700 mbar

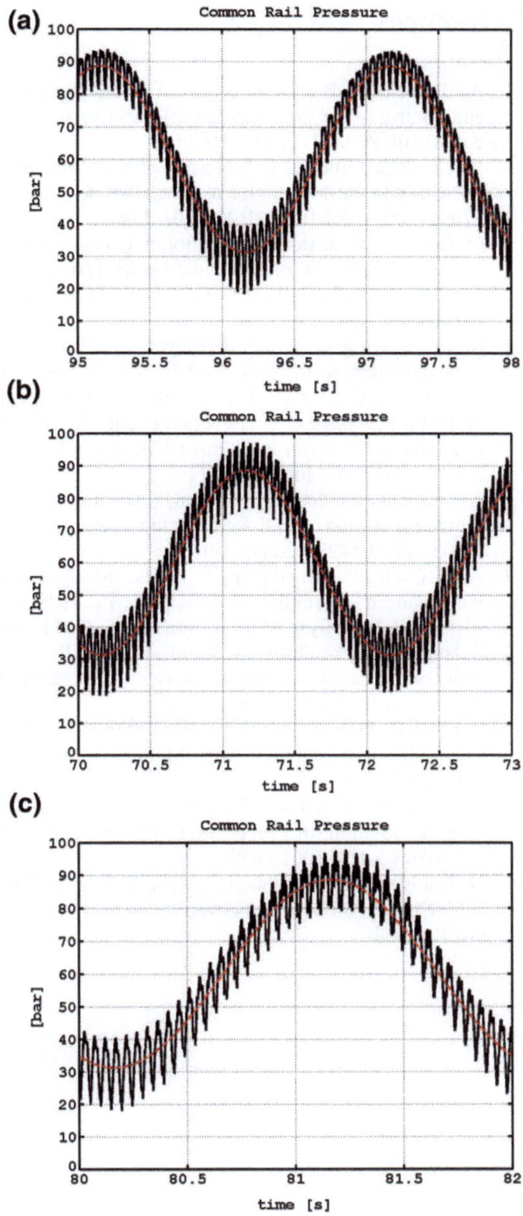

Fig. 4.18 Experimental control results. Total CR pressure $p(t)$ (*black solid line*) and demanded filtered pressure $p_{\mathrm{inj}}(t)$ (*red dashed line*) when engine speed is $N_{\mathrm{e}} = 2500\,\mathrm{rpm}$ and demanded injection pressure is a sinusoidal wave with period 2 s, amplitude 30 bar and bias 60 bar in the case of: **a** *no injections occur*; **b** *single injection* and **c** *double injection* per engine cycle, when the intake manifold pressure $P_i(t)$ changes sinusoidally with period 2 s, amplitude 300 mbar and bias 700 mbar

References

1. A. Balluchi, A. Bicchi, E. Mazzi, A.L. Sangiovanni Vincentelli, G. Serra, Hybrid modelling and control of the common rail injection system. Int. J. Control **80**(11), 1780–1795 (2007). doi:10.1080/00207170701481675
2. W. Chatlatanagulchai, T. Aroonsrisopon, K. Wannatong, Robust common-rail pressure control for a diesel-dual-fuel engine using QFT-based controller. SAE Technical Paper (no. 2009-01-1799) (2009). doi:10.4271/2009-01-1799
3. M. Corno, S.M. Savaresi, R. Scattolini, E. Comignaghi, M. Sofia, A. Palma, E. Sepe, Modelling parameter identification and dynamics analysis of a common rail injection system for gasoline engines, in *Proceedings of 17th IFAC World Congress* Seoul, 2008, doi:10.3182/20080706-5-KR-1001.01434
4. M. Fry, J. King, C. White, A comparison of gasoline direct injection systems and discussion of development techniques. SAE Technical Paper (no. 1999-01-0171) (1999)
5. A. di Gaeta, G. Fiengo, A. Palladino, V. Giglio, A control oriented model of a common-rail system for gasoline direct injection engine, in *Proceedings of Joint 48th IEEE Conference on Decision and Control and 28th Chinese Control Conference (CDC/CCC)*, Shanghai, 2009, doi:10.1109/CDC.2009.5400211
6. A. di Gaeta, L. Glielmo, S. Santini, C.D. Giuseppe, F.D Cristofaro, A. Caraceni, An algorithm for the calibration of wall-wetting model parameters. SAE Technical Paper (no. 2003-01-1054) (2003). doi:10.4271/2003-01-1054
7. A. di Gaeta, U. Montanaro, V. Giglio, Model-based control of the AFR for GDI engines via advanced co-simulation: an approach to reduce the development cycle of engine control systems. J. Dyn. Syst. Meas. Contr. **133**(6) (2011). doi:10.1115/1.4004067
8. S. HaiFeng, Z. YouTong, W. Jun, L. LianDa, Researches of common-rail diesel engine emission control based on cylinder pressure feedback, in *Proceedings of IEEE Vehicle Power and Propulsion Conference*, Harbin, 2008
9. P. Lino, B. Maione, A. Rizzo, Nonlinear modelling and control of a common rail injection system for diesel engines. Appl. Math. Model. **31**(9), 1770–1784 (2007). doi:10.1016/j.apm.2006.06.001
10. M. Tomforde, T. Jeinsch, J. Blath, H.P. Dünow, Modelling of a fuel supply system for model-based calibration. in *Proceedings of 17th IFAC World Congress*, Seoul, 2008, doi:10.3182/20080706-5-KR-1001.2839
11. H. Tomishima, T. Matsumoto, M.Oki, K. Nagata, The advanced diesel common rail system for achieving a good balance between ecology and economy. SAE Tecnical Paper I(no. 2008-28-0017) (2008)

About the Authors

Alessandro di Gaeta received the "Laurea" degree (M.Sc.) in Computer Science Engineering and the Ph.D. degree in Automatic Control from the University of Naples "Federico II", Naples, Italy, in 1999 and 2002, respectively. In 2007, he became Researcher with the Istituto Motori of the National Research Council of Italy. His research interests are mainlyconcerned with the modelling, analysis and control of automotive mechatronic systems and spark ignition engines.

Veniero Giglio got a University Degree cum laude in Mechanical Engineering at University of Naples "Federico II". Since1990 he is with Istituto Motori of the Italian National Research Council, where he is Senior Researcher. His mainexperiences are in the field of Spark Ignition Engine, on both experimental and modeling areas. The most recent interests have been focused on SI engine downsizing by turbocharging and Variable Valve Timing, Gasoline DirectInjection, combustion modeling, investigation on knock and simulation of the related autoignition phenomena, non-intrusive combustion diagnostics by flameionization detection at spark plug electrodes.

Angelo Palladino obtained the laurea cum laude in Computer Science Engineering, in 2006, at the Università degli Studi del Sannio, Italy, where he is currently workingtowards the Post. Doc. degree in the Laboratory of Group of Research in Automation Control Engineering(GRACE). He received the Ph.D. degree in 2009 from the same Institute, in collaboration with Fiat Powertrain Technologies, Napoli, Italy, on automotive field application. His research fields deal with the engine comtrol management improvements, modeling and control of internalcombustion engine, Hardware-Inthe-Loop system simulation and real-time control.

G. Fiengo et al., *Common Rail System for GDI Engines*, SpringerBriefs in Control, Automation and Robotics, DOI: 10.1007/978-1-4471-4468-7, © The Author(s) 2013

About the Authors

Index

G. Fiengo et al., *Common Rail System for GDI Engines*, SpringerBriefs in Control, Automation and Robotics, DOI: 10.1007/978-1-4471-4468-7, © The Author(s) 2013

MIX
Papier aus verantwortungsvollen Quellen
Paper from responsible sources
FSC® C105338

FSC
www.fsc.org

If you have any concerns about our products,
you can contact us on
ProductSafety@springernature.com

In case Publisher is established outside the EU,
the EU authorized representative is:
Springer Nature Customer Service Center GmbH
Europaplatz 3, 69115 Heidelberg, Germany

Printed by Libri Plureos GmbH
in Hamburg, Germany